Praise for *The New Inbox*

"Few email marketers have more experience or go deeper than Simms Jenkins. His latest book once again promises to elevate the dialog and thinking around the most powerful channel in the marketing toolbox, email. Simms Jenkins' new book dives in deeply on this critical channel and provides marketers with state-of-the-art ideas and frameworks to get the most from their programs."

Bill Nussey, CEO, Silverpop

"*The New Inbox* articulates the type of email marketing that will win the hearts and minds of consumers: deeply integrated inbound marketing that leverages email people truly love. The result? You will turn subscribers into customers, customers into evangelists, and your email marketing will become a significantly more powerful and measurable tool in your arsenal thanks to this practical advice from Simms."

Brian Halligan, Co-Founder and CEO, HubSpot and Author of *Inbound Marketing*

"*The New Inbox* is the right book at the right time. Simms Jenkins outlines what works in this new digital world and how savvy marketers can better leverage email marketing. At a time when the inbox has never had stronger ROI for marketers, it has never been more challenging to reach it and engage customers. Anyone who relies on email marketing as a key sales and marketing tool would be wise to follow the advice laid out in this easy-to-read book."

Matt Blumberg, Chairman & CEO of Return Path, Chairman of the Board, Direct Marketing Association and Author, *Sign Me Up*

"As digital marketing has evolved, so has email. Today's email is about driving customer engagement and loyalty, building brand and awareness, and reaching the customer where and when relevant. But getting email right in today's digital marketing environment requires marketers to be much smarter about how they use tools and analytics to deploy email. *The New Inbox* provides valuable insight into how email is evolving in a social and mobile world and how marketers can gain a competitive advantage by transforming their email approach."

Trey Loughran, President, Equifax Personal Information Solutions

---

"Yet again, Simms Jenkins has delivered a work that deserves to be read by not only every marketing practitioner but also any CEO looking to better understand the foundational nature of the email medium—especially in today's increasingly mobile and social world. If your company is ready to stop chasing the latest shiny object and focus on permanently boosting its digital marketing ROI, buy this book!"

Jeffrey K. Rohrs, VP – Marketing Research & Education, ExactTarget & Co-Author, SUBSCRIBERS, FANS & FOLLOWERS Research Series

---

"While there has been an influx of new digital communications channels in recent years, email marketing continues to be a significant driver of our business and a core way that we engage our consumers. Simms Jenkins and his book, *The New Inbox,* lead marketers down the path to being successful in building deeper customer relationships while integrating email with other digital platforms."

Kim Gnatt, Director, Digital Platforms & My Coke Rewards, Coca-Cola North America

---

"Simms shows us why email marketing is not a dying breed. It is, in fact, more important today than ever before as a central component of the digital marketing strategy. *The New Inbox* speaks to how integrating email with social and display programs is still the most effective way to reinforce messaging, amplify the brand and ultimately deliver increased marketing performance."

David S. Williams, Chairman & CEO, Merkle

"*The New Inbox* provides an unbiased look from one the preeminent thought leaders in the industry at one of the most effective ways a brand can connect with its customers in a personal way. This book provides a business perspective on winning and keeping customers that can have tangible outcomes. When people from the supply chain group show up at our new product email planning meetings, you know the results are real, and *The New Inbox* provides the intelligence that can make it happen for any brand."

Michael McCathren, established the Integrated Digital Marketing group for Chick-fil-A, Inc.

"Email growth and engagement have been the most important factors in growing my business and Simms is my go-to email expert. As new things come and go over the years, he has been a thought-leader in what has become the core of online engagement...email. If email might be important to your business in today's increasingly social & mobile world, you won't find many better books to guide your efforts."

David Payne, Co-Founder & CEO, Scoutmob

"*The New Inbox* makes a compelling case for email as a strategic pillar in the digital marketing portfolio. As social media continues to become a "pay to play" model for brands, email is uniquely positioned to drive the bottom-line ROI from paid owned and earned media."

Adam Naide, Executive Director of Social Media Marketing, Cox Communications

---

"In *The New Inbox* Simms Jenkins offers fresh perspectives on the continued viability of email as a necessary and profitable marketing tool.

His insights on integrating email with social and mobile as well as tactics to gain more budget from the C-Suite are useful strategies that every email marketer should embrace."

David Daniels – CEO & Co-Founder, The Relevancy Group, LLC

"I've learned as much about email from Simms as from any other person in the world, and in *The New Inbox*, he takes his common sense approach and puts it out there for everyone. This is a useful, important book about a useful, important marketing tactic written at a historical inflection point that sees profound changes in how email is created, delivered and consumed."

Jay Baer, President, Convince & Convert. Co-author of *The NOW Revolution: 7 Shifts to Make Your Business Faster, Smarter and More Social*

"Email has grown more valuable with each advance in marketing technology along the path to today's ultra-sophisticated, multi-channel powerhouse. Simms Jenkins has been there every step of the way, pushing the envelope to higher levels of customer engagement. Now in his second book to address the state of email marketing, he spotlights what has grown into a wildly expanding, cutting-edge mechanism for building relationships and driving sales."

Stan Rapp, Co-author of *MaxiMarketing* and *BIG IDEAS for Digital Marketers*

"Jenkins proves, through clear and actionable insights, that email marketing is still the cornerstone of today's digital marketing landscape. He provides a fresh, wide-reaching view of email marketing in relation to hotter, high-profile social and mobile channels, breaking the misconception that email's importance is waning. Particularly compelling is his piece on evangelizing to the c-suite, offering tactics for marketers to expose the powerful ROI and connectedness of the email marketing. Excellent for marketers looking to bolster their knowledge and gain a renewed, optimistic perspective on email marketing."

Matt Annerino, VP, Direct Marketing, Live Nation Concerts

"Interested in growing your customer loyalty, retention, sales, and profits? Buy this book! Blow away your competition by learning from one of marketing's thought leaders how email marketing can leverage your mobile, social, digital, and other marketing efforts."

Ken Bernhardt, Regents professor of Marketing Emeritus, Robinson College of Business, Georgia State University

---

"Simms is certainly one of the brightest minds in digital marketing. Whether I am having breakfast with him at the local Waffle House or reading his latest book, *The New Inbox,* I am amazed at his vision for the future of email and digital marketing."

Allen Nance, President & CEO, WhatCounts

"In *The New Inbox,* Simms Jenkins highlights the central role that email continues to play in a world that's only getting more social and real-time. Brands, Agencies and Practitioners who want to understand how email marketing and social media can play well together should read this book and put the case studies Jenkins' shares into practice for their own businesses. Put email at the center of your social CRM strategy and you will succeed."

Dave Hendricks, COO and CMO, LiveIntent

"A lot has changed in the Email Marketing world since 2008 when Simms Jenkins' last book, *The Truth About Email Marketing,* came out: The dramatic rise in Social Media as a marketing platform and the near ubiquity of mobile smartphones have changed the email marketing landscape forever. Great news then that Simms has written a new treatise, *The New Inbox,* to get us all up to speed. This is the 'real stuff' written by someone who is in the trenches everyday. *The New Inbox* is the clear and practical guide that every email marketer needs to navigate today's digital marketing environment brought to you by one of the most respected and knowledgeable people in the industry."

Bill McCloskey, Founder, Only Influencers and eDataSource.com

---

"The Internet keeps changing and reinventing itself. We go through phases of extreme innovation, then phases of extreme disruption, and then we start all over again in a whole new direction. It can be maddening or fun, depending on your point of view. With each revolution, we get a new generation of innovators with new technology, new ideas and new perspective. And each new generation needs to communicate well (that never changes). Thing is, communication is hard! It can be maddening or fun, depending on your point of view. And since email is the foundation for communication on the Internet, it seems each new wave of bright young innovators needs a fresh new guide to help them understand the inbox. As always, Simms delivers."

Ben Chestnut, CEO, MailChimp

"Email has evolved from a disregarded workhorse to a versatile powerhouse of digital marketing. Its role continues to expand to support marketing automation, the mobile revolution, social strategies, and advanced acquisition and retention approaches. Sitting at the forefront of email marketing, Simms Jenkins is ideally placed to guide email marketers through the associated opportunities and challenges. He builds a bridge between email's traditional role and its growing position as a driver of integration, explaining how marketers can and must adapt best practices to the requirements of the new digital environment as they look to exploit the huge potential of *The New Inbox*."

Mark Brownlow, Publisher, Email Marketing Reports

To Kelly, Sam, Cal, Sloane, Mom and Dad for inspiration and encouragement.

The New Inbox

Simms Jenkins

ISBN: 978-0-9890518-0-4

Portions of this book first appeared as columns written by Simms Jenkins on ClickZ (www.clickz.com) and are used here with permission.

BrightWave Marketing
Jessica Johnston Higgins, Cover Designs
Julia Bray, Project Coordinator

ClickZ
Caitlin Rossman, Copy Editor
Dawn Cavalieri, Creative Manager

Table of Contents

Forewords

Instant coffee didn't kill brewed coffee. Popcorn chicken didn't kill chicken. Television didn't kill radio. The Web didn't kill print (yet). And social media won't kill email.

Sure, social media is exciting and interesting and visceral. And of course it merits attention when Facebook has one billion users, and more online time is spent engaging in social media than in any other activity. But the rise of social media hasn't diminished the impact of email, and in fact it may have increased it. After all, what do you need to become a member of any social network? An email address. And what does every social network use as a "hot trigger" to get you to come back to their site to see who's said what about whom? Email.

One of the great fallacies of social media is that it's primarily a customer acquisition vehicle. The reality is that most companies are preaching to the choir in social media. We "like" what we like, and the consumers that friend, follow and subscribe to your brand's social missives are disproportionately likely to be existing or former customers. Thus, the most logical role of social media in the communications ecosystem is to keep the brand top-of-mind by providing ongoing, relevant information, with a smattering of entertainment to keep things lively. Sound familiar? Other than in the most transactional circumstances ("your boarding pass is attached"), the role of email is strikingly similar to that of social media.

Thus we uncover a great truism of modern marketing: Email is Madonna; the original but fading out of the public consciousness after many years in the spotlight. Social media is Lady Gaga; essentially the same thing, but with a fresh coat of paint that grabs all today's headlines.

It's not a one or the other proposition, as today's smart marketers need both Madonna and Gaga. And in fact, most companies have no business getting serious about social media until and unless they have an outstanding email program. Why dive into the deep water of a synchronous and very public communication platform if you haven't mastered an asynchronous and less public modality that gives you a much larger margin for error?

Simms Jenkins gets this principle in every way it can be gotten. He was doing email before email was cool, and now that it's less cool than ever, he's helping marketers understand that headlines and hype do not equal effectiveness. Maybe you get it, too, as more marketers are planning to increase their email budgets this year than even, that's right, social media.

I've learned as much about email from Simms as from any other person in the world, and in *The New Inbox*, he takes his common sense approach and puts it out there for everyone. This is a useful, important book about a useful, important marketing tactic written at a historical inflection point that see profound changes in how email is created, delivered and consumed.

Ignore these instructions at your peril, and make sure you have a highlighter. This is a book you'll want to highlight and refer back to again and again.

Read on and see for yourself.

Jay Baer
President, Convince & Convert
Co-author of *The NOW Revolution: 7 Shifts to Make Your Business Faster, Smarter and More Social*

We have come a long way from the early days of spray-and-pray email messaging

Email has grown more valuable with each advance in marketing technology along the path to today's ultra-sophisticated, multi-channel powerhouse. Simms Jenkins has been there every step of the way, pushing the envelope to higher levels of customer engagement. Now in his second book to address the state of email marketing, he spotlights what has grown into a wildly expanding, cutting-edge mechanism for building relationships and driving sales.

There are pundits who tell us email is about to fade away with the rise of social and mobile; Simms Jenkins shows why just the opposite is true. Email remains at the heart of digital marketing. It's here to stay as marketers adapt wisely to the many opportunities new technology makes possible

The inbox is not going away. What is essential for digital marketers to understand is the transformation taking place in reaching out over new channels to consumers who may choose to respond or not respond for a wide range of factors. With each gain in getting closer to the customer and with each new channel opening, you must recalibrate the purpose and the size of your email marketing investment.

In *The New Inbox*, Simms Jenkins dismisses the false notion of email dying because teenagers are taking their messaging elsewhere. In a world where the typical smartphone user spends almost half his time on email, the inbox isn't going away. It just may look much different.

Email remains the number one way to reach people in the online world. There's nothing in sight to change that reality any time soon. With a return of $40 for every dollar invested, according to DMA research, it more than pays its way. Now with the dawn of a new email creative era, you can expect that return to grow even greater.

Stan Rapp
Co-author of *MaxiMarketing* and
BIG IDEAS for Digital Marketers

Preface

The second time of any challenging and rewarding journey is always supposed to be easier than the first. The process for this book (my second after The Truth About Email Marketing), has been less intimidating and foreign, but I am not sure if it has been much easier. Nevertheless, I have enjoyed it and feel a bit wiser and am looking forward to the fun part – that is when people get their hands on the book and hopefully learn, test, improve and succeed.

I would like to thank the team at ClickZ and Incisive Media for making this happen. Mike Grehan, Caitlin Rossman and Melanie White have been great partners in this effort.

A great deal of patience, support and expertise has come from my team at BrightWave, the best group of email professionals in the world. The same goes for our many clients who trust us to elevate their email programs. As BrightWave celebrates our 10th anniversary, we know how fortunate we are to have great partners.

The email marketing industry is an amazing and unique one. Many people have been in this line of work for over a decade (which is uncommon in the digital world) and have a lot of common bonds and fascinating war stories. Email, of course, doesn't get the respect that it should and that seems to form an even tighter bond for many of the hard working people in our space. It is wonderful encouragement to see many smart and talented professionals work so diligently to move our industry forward on many fronts. I hope this book helps in just a small way to make them feel more valued. A special thanks to the all-star cast I asked to provide feedback or a quote about the book.

As always, my family is what inspires me and keeps my feet on the ground. Thanks to my parents for being cheerleaders for everything that I have tackled in life – big and small. My sister Bryn and brother-in-law Craig are fantastic supporters and friends, and I wish their family lived closer! The same goes to my grandparents who continue to amaze me. My in-laws have always been a source of wisdom and guidance and I appreciate that. Thanks Laura, Rosanne and Wayne.

My wife, Kelly, has me in awe for many reasons and despite her many endeavors that she is constantly juggling remains my number one supporter. It's hard to believe what has happened in the 10 years since BrightWave was launched in our little house on Lake Avenue. Thank you for being the best wife, mother and partner anyone could ask for.

Last but not least, my three children, Sloane, Cal and Sam, are the motivation for so much of what I do and provide the joy and balance required for a complete life. While I don't think they understand what daddy does for a living, hopefully this book paints a better picture and provides a little more of a tangible legacy.

Introduction

"It has become appallingly obvious that our technology has exceeded our humanity."
— ALBERT EINSTEIN, PHYSICIST

"Man is still the most extraordinary computer of all."
— JOHN F. KENNEDY, 35TH PRESIDENT OF THE UNITED STATES

I launched a little one man business out of the humble guest room in the tiny house that my wife and I bought when we were first married. Before I started what became BrightWave Marketing, my wife and I both lost great jobs at technology firms and certainly had the wind knocked out of our sails. We were flying high, like many others, during the dot com era and fully expected the good times and seemingly boundless opportunities to continue. Then it came crashing down, and it hurt. More than one wise person told me starting at the bottom as a newlywed and entrepreneur was a great thing. There was nowhere to go but up!

There is certainly a lot of wisdom in that advice and it mirrors email marketing on many fronts. Email marketing was just a "thing." It was a list. A blast. A cheap and easy way to broadcast a message to nameless people that could be customers. From its humble roots, it struggled on many fronts but is poised to continue as the central and most reliable way to communicate with a brand's customers and prospects.

As my firm celebrates its 10th anniversary, we are thankful and maybe a bit lucky that email is the old fashioned digital platform that gets results. Results are what matter for most marketers and executives. Email gets social media fans and followers. Email gets new customers. Email keeps customers. Email drives revenue and generates a significant return on investment. Emails work when relevant content and valuable offers get delivered. Email works when permission is granted. Email is what customers want.

The last three are the real key – it doesn't matter what new technology and communities develop, email will always have a role in consumers' lives as long as permission is present and value is what shows up in the inbox. Email is the digital glue, the hub, the linchpin. I could go on but email marketers need not be threatened by emerging communications channels and networks. 93% of US online consumers are email subscribers, receiving at least one permission-based email per day, according to ExactTarget.

The email marketing world has withstood the storm of amazing new digital tools like Facebook, the iPad, text messaging, Twitter and more. All of these things work well with email which is the beauty of email's quarterback position on the marketing field. How do most brands generate a following on Facebook? Email. Have you seen how beautiful many emails look on the iPad? Text messaging and SMS have not taken off in the corporate marketing sense but forward thinking brands leverage this familiar technology to acquire new email subscribers. Twitter uses its plentiful content to get you back and actually tweeting.

The New Inbox isn't about defending email marketing or articulating why it's better than social media or the latest and greatest internet fad. It's about embracing the new way consumers interact with brands and adapting to how email fits in this new environment. It's not a computer sending a message to a list. It's a person behind a brand sending a message to another person who cares about the brand. While marketers rely on technology, we thrive when the human side is what is demonstrated to the end user which is (at least for the near future) a human.

The new inbox of consumers literally looks different than ever before. Much of the same fundamentals still matter though and this book hopes to cover these areas as well as lead you to where email is headed. My hope is this book serves as part road map and part stories from the trenches. Whatever you make of it, I hope it brings you success.

As a reader, I hope you feel this book is more like a conversation and less like a lecture or some random dude telling you how to be smarter or better at your job. As you will see (if you don't already know), I am a big fan of lists, bullets and storytelling. My desire is this book is easy to read, relevant, insightful and spurs lots of testing, new ideas and success.

Finally, feel free to visit my book website at www.TheNewInboxBook.com or like the book on Facebook at http://www.facebook.com/TheNewInbox. Hopefully, the conversations and learnings can be mutual and continue on these locations. Thank you for your time and interest.

Simms Jenkins
March 2013, Atlanta, GA

"The most important single central fact about a free market is that no exchange takes place unless both parties benefit."
MILTON FRIEDMAN

CHAPTER 1

The Big Picture for the New Inbox

Email has gone from an inexpensive weapon in the marketing toolkit to the indispensable digital hub for all brands. It is still the most targeted, measurable and cost effective way to reach your customers and prospects but how you do so has changed in dramatic fashion. It is still all about the power of permission; however, smartphones, tablets, Facebook, Foursquare, Twitter and more have radically altered our communication landscape. Email continues to be the glue that binds all marketing channels but marketers must adapt and evolve as the email channel has or risk being left behind. Best practices are often stagnant and stale – now is the time to understand the new inbox and ensure you find your way in to it.

Inbox Rising: Email Emerges as the Dominant Digital Communication Channel

"56% of marketing executives planned to increase spend on Email Marketing in 2013, 52% on Social Media."

"Email is more relevant today than ever before as consumption continues to grow on more platforms."

Yes, we're moving to an integrated digital world (well, some of us are), but email has emerged as a more dominant channel than many of us even realized. More mobile, more social, and more forward-thinking avenues to talk to our customers and prospects are emerging every day, yet email's role was not fully cemented as the key platform to bridge the disparate worlds together.

Above I referenced two studies (StrongMail's 2013 Marketing Trends Survey[i] and a quote from Return Path's "Mobile, Webmail, Desktops: Where Are We Viewing Email Now?"[ii] respectively) that illustrate just how email is not only transforming but deeply entrenching itself as the number one way to reach people in the online world.

So where and how is email moving the needle for digital marketers? Let's look at the seven big areas where email can support my claim as the digital marketing heavyweight to beat in this new era of mobile and social marketing: Acquisition, Awareness, Location-Based, Loyalty, Retention, Revenue and Social.

1. *Acquisition*

I won't get into the whole list-renting world (at least in this chapter), but email still works for customer acquisition. Thirty-two percent of marketing executives say it's among their most important email initiatives. Search, display, mobile, social, and offline media can drive not only sales but also lead in the form of email subscribers. Email can then extend that media buy and reach by efficiently monetizing them as well. Successful companies are realizing that email is the most profitable channel they have and that it provides a longer tail to their other marketing investments if they can convert these "eyeballs" to email subscribers where they likely have a great shot at being monetized later.

2. *Awareness*

Email as a branding channel? What? I have many clients who view that just reaching their millions of subscribers in the inbox on an ongoing

---

i http://www.strongmail.com/pdf/SM_Trends2012.pdf

ii http://www.businesswire.com/news/home/20111206005201/en/Return-Path-Report-Shows-Mobile-Platforms%E2%80%99-Growing basis as a win. Notice I didn't mention opens, clicks, or conversions. Just an email being noticed (maybe in the from line or a killer subject line that does its job and lets the subscriber move on). Surely, many consumers have experienced this on Black Friday or Cyber Monday.

Think about it: you are in the mall and you glance at your smartphone while in line (because that is what we do). You notice a 50% off email from the department store that happens to be 100 feet from where you are. You don't need to read the email – it worked. Just a glance of the compelling offer was relevant enough to send you on your way. From an attribution standpoint, that is a potential nightmare but also a win for any marketer.

An email doesn't have to be clicked upon (or counted as an open) to work and that in and of itself is powerful.

3. *Location-Based*

Being able to know where your customers are is cool, especially when they're at your location and you can reward them or cross-promote something (for example, leveraging Twitter on Foursquare to drive email subscriptions). Reading an email while you're on the hunt is even more powerful from a sales perspective.

Adapting email campaigns to smartphones and tablets will no doubt correlate to your ability to convert these subscribers on the fly especially since this is the biggest trend of email consumption (more on that later). Whether sending device specific emails or leveraging real time located based data to drive subscription, email will need to tap the power of understanding where your subscribers are to hit the next level.

4. *Loyalty*

Our agency has several restaurant clients that truly view their email program as a loyalty program (without the annoying punch cards). That is a great way to think about your program. Don't think about it as a list or file (that term makes me shudder). These are real people and even better they gave you permission to email them. How great of a first step for loyalty is that? You must live up to the promise though which can be the challenging part.

Email drives frequency, strengthens ties to your brand, and should increase the lifetime value of any customer. That's also why the downside is so strong when it comes to abusing your subscribers. Loyalty can erode with too many valueless emails.

5. Retention

Not only is it less expensive to retain customers than acquire new ones, but email provides us with the best method of doing so. Just about anyone engaging in email marketing is doing this, perhaps without even knowing it. Many marketers refer to this as mailing to their house list, sometimes in a negative way which is counter-intuitive since these subscribers actually want to be on your list. They have either opted in to your email program or purchased from you (often both). This is your best source of email power in the short and long term.

6. Revenue

We know this is the main reason we all get bombarded over the holidays: email sells stuff! However you may track revenue, email is likely contributing to your bottom line in some shape or fashion. If not, you are missing out given the huge potential here for a return on investment. The Direct Marketing Association's latest numbers (from 2012) say email returns almost $40 for every dollar spent. Not bad, if you can get it, right?

7. Social

This has been discussed more in the past few years than any other email topic. You can ask any social media manager how they've built or plan to build a community on any social platform. They'll tell you email is a key tie in to social, but there's plenty more to come in the future. Leveraging a more "social" personality in many email campaigns and sharing worthy content will make your email blend more seamlessly with social tactics and possibly vice versa.

The New Digital Inbox – Is Your Email Program Ready?

The ways in which we access and read email – via smartphones, tablets, and social networks, for example – are evolving rapidly. Most email applications are not very forward thinking, using outdated templates and former best practices. Today, email may not be a standalone digital conduit for brands to pipe through deals, offers, and newsletters. The game-changing opportunity is at our front door.

Trends in Mobile

While things like daily deals and social networks are creating a new type of inbox, mobile devices and technologies are also shaping the ways in which people digest content. Smartphone and tablet users are being conditioned to access a unified inbox for all digital messaging communication. The home screens of most smartphones are becoming the starting point for decision-making when a new message arrives. Facebook updates, tweets, email, and more are all arriving on the home screen, with little discernable difference.

Honeycomb, Google's tablet version of Android, is pushing the centralized notification even further with its notification icon bar and enhanced widgets. Apple's mobile operating system, iOS, is continuing to centralize the digital messaging stream as well; iOS 5 included a notification center that goes beyond the push notices and icons in iOS 4. This means that texts, social requests, email, and more will all be in once place, and the user may not even be aware of the distinctions. Your competitors' tweets and SMS messages may be right next to your email campaign.

5 Tips for Creating Better Emails

In light of these trends, what do you need to know and do in order to transition and succeed?

1. *From and subject lines are growing in importance.* They decide whether your email is immediately deleted, read, or saved for another time. Your subject line is now competing side-by-side with tweets and posts.

2. *Brevity wins.* It doesn't matter whether it's an email to your boss or to your five million subscribers – the future of digital messaging is about succinct and valuable content. If you have a killer offer and wait until the fifth paragraph to provide the link and call to action, forget about it. You just lost millions in potential sales because you failed to adapt to what people want and how they read messages. Remember that smartphone users may only see half (or less) of the content of your email when reading on their mobile device versus their desktop.

3. *Use alt text and pre-headers.* Sure, killer creative still has a place in the next generation of email, but don't force your customer to click on "view image" or enlarge the email to see the entire HTML version, as that is an added step for them and another to delete. This means that the intro copy and the text that you provide when images are not displayed need to be snappy, to the point, and able to sell the email all by themselves.

4. *Move beyond "looks good on mobile."* Most email programs design email for the desktop while ensuring that the email looks good on mobile devices, making them "degrade gracefully." Is this the standard? That we design email for mobile devices that just doesn't look bad? We should design mobile-enabled email that is not only optimized for the device, but also has content relevant to the mobile user. It's not good enough to just make an email passable for the smartphone user; the campaign should be optimized for the device in both content and layout regardless of whether that user opens the email on their smartphone, tablet, or desktop.

5. *Create a deeper relevance.* Relevance has been a hallmark of email marketing for quite some time, but in the age of the new inbox, relevance is even more important. Give me the right offer/content at the right time, or else. Think about users dismissing a text from their boss while dining with their spouse, or that same self-promotional tweet from the latest social media charlatan. It's about basics: the right offer and content when the user wants it. Otherwise, you are just a deleted message, no matter how you sit on the device.

The Email Marketing Trends That Matter For the Future of Email

Now let's look at emerging trends that have staying power. I mean real ones that will likely continue to surface and cause each and every email marketer to adapt in a meaningful way.

Rule #1. Major Creative Overhaul

When did email creative get sexy? Well, it is now and it is often what gets email shared, clicked and a subscriber to buy something from you. That may sound like a boring start from a trend standpoint but we are seeing a renaissance (or is it the dawn of the email creative era?) of sorts where marketers care more about the impact their email creative makes. This is one of the much needed departures from email's direct marketing heritage. No more form letters or copywriters dominating the email process. The creative director is a new force in the email game, and he or she views email as a new tapestry rather than tight box to fit in.

For those smart, pragmatic readers looking to hit singles and doubles and taking notes of low hanging fruit, start here. One way to do that is to make wholesale changes to your (often) stale email creative and messaging. Ask yourself: does your creative look like it was built pre-iPhone? Has it been truly touched and refreshed to account for your brand's evolution and your subscribers' inbox viewing changes?

Whether it is an automated message that IT created and hasn't updated in years or your key revenue-driving promotional templates, these messages must be refreshed frequently and many times must be thrown out the window in exchange for new and more relevant creative. Strategic creative changes can have the most dramatic and timeliest impact on your email program.

Rule #2. Mobile

First let's connect to the previous item. The best thing that's happened to email marketers in the past five years? Apple's iPhone (and its competitors, followers and clones) thankfully beat out less HTML email-friendly smartphone makers like RIM's BlackBerry. This has led email to be the number one activity that consumers perform on their omnipresent smartphones. So those beautiful emails will render much nicer on the iPhone than a draconian device.

But this goes beyond email creative and toward rethinking how we communicate to our subscribers. We need more right-time, right-place messaging (and remember that 78 percent of United States email users will access their emails with a mobile device by 2017, according to ForresterResearch). We now need to not only drool at this prospect but to plan how to trigger an email to a subscriber after they check in at one of your locations, acquire a new subscriber through a new experience like an app or social network, or learn the best way to serve up the right email coupon so your offline staff can handle and track it in the most efficient manner.

If you need more data to ponder, consider this: *mobile purchasing decisions are most influenced by emails from companies (71%)* only surpassed by the influence of Friends (87%) (Adobe "2013 Digital Publishing Report: Retail Apps & Buying Habits")

Rule #3. Integration

Both from an internal and external perspective, email will become more in sync with what's going on within your company and outside of it. This means coordinating deeper teamwork and education with the groups that power email's wingmen (search and social) to e-commerce, technology, and offline efforts. If your email program lives in an isolated existence, you must seek a way to break out of this silo. You will be doing yourself and your company a major service.

Rule #4. People

The unsung heroes of any marketing department (it's not the media or technology, it's the people!), email folks toil in near obscurity yet are the ones making or saving their employee a substantial amount of money. It's not the email "machine" driving millions of leads and dollars but the people and partners behind (and in front of) any technology. With the economy showing a modest recovery and email's proven ROI, serious digital marketers will stock up to find the right teams to help move their email program from one that manages and delivers emails to a versatile and strategic one that becomes adept at moving the business forward, not just the campaign message.

Digital's Great Teenage Misunderstanding

A lot of data has come out on how consumers' digital habits are shifting. None has been as controversial or frankly inaccurate as comScore's assessment[iii] of web-based email usage for teenagers.

ClickZ.com ran an article, "Email Usage Plummets as Teens Turn to Mobile, Social Networking,"[iv] about the report. To quote, "most noteworthy was the shift in email usage, particularly among young people. Total web-based email use was down eight percent last year, led by a walloping 59 percent drop among 12 to 17 year olds."

I must reemphasize, the data is only for web-based email usage (think Hotmail, Yahoo Mail, Gmail, etc.) and that's an important distinction. A decline is a decline, but this certainly doesn't fully cover how email is consumed in today's digital world, especially as mobile email consumption seems to rise every week (one late 2012 study from Litmus had 43% of all email being read on mobile devices).

No less an authority on teenagers' digital consumption than Mark Zuckerberg offered this at the Facebook Messaging announcement:

> "High school kids don't use email, they use SMS a lot. People want lighter weight things like SMS and IM to message each other."

It's hard to argue with this thesis. There are two significant issues that must be added to the conversation though:

---

iii http://www.comscore.com/Insights/Presentations_and_Whitepapers/2011/2010_US_Digital_Year_in_Review

iv http://www.clickz.com/clickz/news/2025027/e-mail-usage-plummets-teens-mobile-social-networking

Mobile's impact: The typical smartphone user spends almost half of her time on email. This makes comScore's metrics marginal since it evaluated only web-based email usage.

As e-Dialog's former CEO John Rizzi thoughtfully points out on a blog post[v] in response to this data:

> "In the 18-24 age group, unique visits increased 9%, while time spent decreased 10%. To me this points to the increasing use of mobile to triage inboxes on the go, and the desktop inbox being used to access specific emails and perform tasks like getting a code for a sale, or composing an email reply that would be too onerous on a mobile phone. In fact, comScore found that 30% of respondents are viewing email on their mobile phone, a 36% increase from 2009, and those using mobile email daily increased 40% on average."

Pew Internet evaluated[vi] how Internet users of different age groups spent their time online. Guess what? Even 90 to 100 percent of millennials (ages 18-33) used email. As you can see in the chart, below, email was the top activity across all age groups.

Teenagers become adults: I may not win any scientific breakthrough awards for this statement, but people are missing the boat on this issue. Teenagers don't remain teenagers forever (thankfully)! What happens when a teenager becomes an adult in the workplace? Not only do they dress, speak, and act differently – they use different approaches to communicate, as well.

The first thing a new employee typically gets is...an email address. And guess what? They use it, even if they have been reliant on social, IM, and texting for their primary communication channels. They will correspond for work via email and opt in to emails from their favorite brands (including brands that they certainly did not like as a teenager). They will also "Like" their favorite companies on Facebook, follow them on

---

v http://www.therelevantmarketer.com/2011/02/e-mail-use-isnt-declining-its-just-shifting-from-web-to-mobile.html#tp

vi http://pewinternet.org/~/media//Files/Reports/2010/PIP_Generations_and_Tech10.pdf

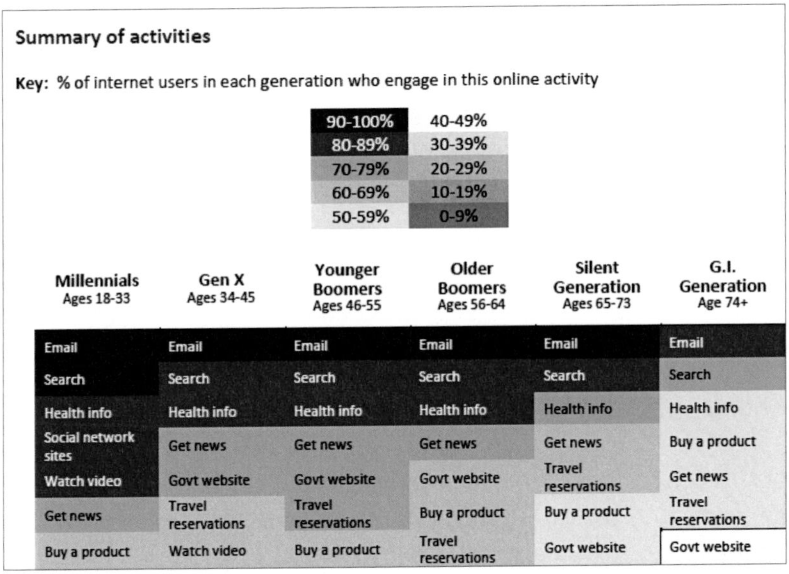

Summary of activities

Key: % of internet users in each generation who engage in this online activity

90-100%	40-49%
80-89%	30-39%
70-79%	20-29%
60-69%	10-19%
50-59%	0-9%

Millennials Ages 18-33	Gen X Ages 34-45	Younger Boomers Ages 46-55	Older Boomers Ages 56-64	Silent Generation Ages 65-73	G.I. Generation Age 74+
Email	Email	Email	Email	Email	Email
Search	Search	Search	Search	Search	Search
Health info	Health info	Health info	Health info	Health info	Health info
Social network sites	Get news	Get news	Get news	Get news	Buy a product
Watch video	Govt website	Govt website	Govt website	Travel reservations	Get news
Get news	Travel reservations	Travel reservations	Buy a product	Buy a product	Travel reservations
Buy a product	Watch video	Buy a product	Travel reservations	Govt website	Govt website

Source: Pew Research Center report, "Generation Online in 2010"

Twitter, and opt in to SMS offers as well. They will also expect different value and information in each of these channels.

I culled some great and simple statements from the Twittersphere that sum up this dangerous notion of "email is dying because of the teenager." For the "email is dead/will die" crowd, I offer up Exhibits A, B, and C.

> Kids don't drive. Therefore, cars will soon cease to exist. #emailisdeadanalogy - @MartinLieberman[vii]

> Most teens don't drink red wine, red wine is probably going away #emailisdeadanalogy - @LorenMcDonald[viii]

> Most teens don't have kids, there will be no children in the future #emailisdeadanalogy - @mostew[ix]

The inbox won't be gone in the next 10 years. However, it will look a lot different. That means email marketers should ensure they innovate

---

vii http://twitter.com/martinlieberman

viii http://twitter.com/lorenmcdonald

ix http://twitter.com/#!/mostew

and deliver compelling value to the inbox regardless of a subscriber's age and where she reads her emails.

More importantly, the inbox itself will be displayed and accessed differently. Gone are the days where people limit themselves to five utilities on their desktops (i.e. web browser, Outlook or email client, Excel, Word and PowerPoint). Today, people literally have countless ways to interact with their favorite brands and friends.

Changing Consumer Perceptions of the Email Channel

At a recent Email Evolution Conference[x] hosted by the DMA's Email Experience Council, I saw Ryan Phelan, an executive at Acxiom, present a keynote presentation along with Shar VanBoskirk of Forrester Research. He provided some compelling new research and "consumer-on-the-street interviews" that move the email conversation away from "us" and back to the end users.

I asked Ryan his thoughts on the three most interesting and timely areas of email marketing. We discussed the changing views of consumers and how digital marketers need to adapt. They are an important window into how email marketing needs to change since the consumers view of the channel certainly is.

Simms Jenkins: From your research, you found that 95 percent of consumers said they sign up for emails because of discounts. But why do people use email to get these discounts when these offers can often be found elsewhere? Why don't they get these from other channels like social, web, or mobile? What makes email special and unique here? Why is there more trust in email than even the brand's own websites?

Ryan Phelan: I don't think this is an either/or thing. Many people are likely looking for discounts multiple ways – signing up for email to get them and also going to social, the web, and text for them.

---

x http://www.the-dma.org/conferences/emailevolution/index.shtml

Customers sign up for email or any communication based on the type of relationship that they want to have with the company. Consumers are very familiar with email, since it is also how they communicate with their friends, family, and co-workers. And for discounts, email allows the subscriber to get discounts delivered directly to them, where they can manage and access them when and how they like. Add mobile email to the equation and the subscriber's access to discounts increases even more.

Those who choose to get their discounts via social or mobile or brand websites are no less valuable than those who choose email. They just have different preferences, which we should acknowledge.

Simms Jenkins: You mentioned that marketers need to avoid taking advantage of the trust subscribers give us. What are some common ways that email marketers seem to be taking advantage of their subscribers, even unknowingly?

Ryan Phelan: I believe that almost everything email marketers put out there for their brand either builds or erodes subscriber trust. Very little is received neutrally.

When marketers communicate in a one-message-fits-all way that fails to acknowledge each subscriber's unique experience with a brand or product, trust is eroded. Similarly, if email frequency is too high for the subscriber and they're not given the ability to modulate that frequency, trust can be eroded and brand credibility can be affected.

You can look at both of these examples and point to a lack of segmentation or frequency caps, and that's usually part of the problem. But these are also symptoms of a larger issue that I believe is the real reason we're taking advantage of subscriber trust: the marketer makes assumptions about what the consumer wants, either because they're placing their objectives ahead of subscribers' interests or because the diligence hasn't been done to listen to subscribers and understand their needs with greater confidence.

Simms Jenkins: In 2006, 77 percent of subscribers said they received too many emails and promotions. However, in 2010, that percentage shrunk to 49 percent. Why? What changed on the marketing front/and or subscriber front?

Ryan Phelan: I think this is because email has changed and because subscribers have changed. Back in 2006, many of the emails brands were sending out were basically copies of their website pages, or in many cases, newsletters.

Just a couple years later, when the economy worsened in the U.S., marketers began relying on email more for sending discounts and offers. As we've already discussed, this is a behavior that subscribers have become accustomed to and it's something that they feel is valuable. Additionally, subscribers are becoming better educated about managing their email volume and expecting ISPs to help with some of that, as well.

Over the last 18 months, we've seen the ISPs moving toward relevance-based deliverability. Google, for example, may consider an email spam if there is not engagement by the user. This challenges marketers to come in line with the new "rules of engagement" and to recognize that a relationship-based program drives higher results. As they see higher results with triggered messaging, they are starting to move toward a regular cadence of segmented emails. Sure, this is moving slowly, but consumer expectations for it are growing.

Simms Jenkins: Relevance is a term a lot of marketers disdain mainly because so-called email experts are usually harping on it and because it can be seen as vague - yet it was the number one reason for people to unsubscribe. What do you make of that?

Ryan Phelan: Simms, I believe that this is one interesting outcome of years of consumer experience with search engines and the web. Consumers have become familiar with – and have gained value from – delivery of content that reflects their individual needs and interests. They've learned to expect that from their search results, from the products they see on the web, and from what's in their inbox.

It makes perfect sense for consumers to call out marketers for lack of email relevance, particularly today when they are more empowered because of access to social media. What's more remarkable is that we are surprised when consumers call us out for it!

Simms Jenkins: You take a look at some old-school and what I think can be irrelevant best practices (such as asking people to add you to their delivery book/safe sender). For example, you found 70 percent of people don't add marketers to their delivery address book. What does that tell you about subscribers and their understandings and actions as they relate to the technical side of email?

Ryan Phelan: I'd encourage us to focus on the percentages of people who do it! When we consider that subscribers are looking for and demanding relevant content, and that they are protecting their inbox from getting flooded, the 30 percent of people who are adding a company to their address book are likely brand loyalists or advocates. As for their understanding of this, I think they know what the pre-header functionality does, they are just making choices about whether to click on this or not.

Simms Jenkins: Forrester said most email programs receive a failing grade, according to its latest review. Why can't we get smarter and better given that email is a relatively mature channel compared to their digital communications and messaging channels? Will we ever "pass" as a collective industry?

Ryan Phelan: Simms, I think we are getting smarter, we're just not consistent. There are two key things driving this: first is the amount of churn in digital marketing. College grads often work their way into digital marketing by doing email and after a bit of experience they move on, taking their knowledge with them. So as providers, we are always educating and helping marketers learn, often when they don't have ready access to learnings from their predecessors.

The workload of marketers is the second factor that contributes to program failure. Email marketers wear many hats and many are asked to be tactical without the time or space to be strategic. Most program failure comes from a failure to start with a clear strategy that rolls up under higher level business goals and objectives.

Both factors impact our ability to drive toward a "pass," but it has taken and continues to take a lot of work. That's where I think the strategist role is essential. It provides a more holistic view of the program, plus helps ensure continuity over time and across channels.

Simms Jenkins: What is your advice for email marketers trying to adapt to the fact that many of their subscribers are reading their emails on smartphones and tablets?

Ryan Phelan: One of the statistics from our study was that 69.7 percent of consumers deleted and 18 percent unsubscribed from emails that looked bad on their mobile device. This statistic was also backed up in the street interviews that we conducted. Considering the rate of mobile adoption, it's a stat that should have marketers sprinting to optimize their emails for mobile as quickly as possible - especially since most ESPs offer this functionality. Email marketers need to put their brand everywhere their subscribers are consuming information. More and more, that's on their mobile device.

Simms Jenkins: Ryan, any parting words or takeaways from the EEC event and conversations you have had with marketers of late?

Ryan Phelan: This is a fantastic industry that is truly ready to help marketers who want to be successful. I challenge marketers to pay attention – both to how their customers behave, and to what they tell you they want and don't want from the relationship you have with them.

The Changing Landscape of Email Investment

I've been in the email business for over 10 years, and while not much has changed in our space, we have seen a pretty dramatic shift in what was previously a simple and straightforward budget line item. A lot of the changes are due to an emergence of peripheral email developments (like smartphone consumption and social media adoption) and specialized email-focused companies, as well as the more macro issue of email marketers getting more sophisticated and realizing a good email program absolutely requires more than just a delivery platform.

Let's look back at a typical email marketing budget from five, seven, or 10 years ago.

Delivery/Software

This often was 100 percent of the budget of many marketers. Selecting an email service provider (ESP) was often seen as the one and only decision and requirement for success.

Deliverability/Rendering

Companies like Return Path and Pivotal Veracity filled a void that email marketers continue to find needed – how do they ensure emails get into their subscribers' inboxes and measure their effectiveness?

Acquisition

A plethora of companies would gladly sell, rent, and help find new subscribers, often regardless of whether they provided permission or even knew who your company was. In an effort to grow small programs, this area of the budget often was the most focused on, yet most off the target. In many infamous charts from research providers, this line item often fell in the email advertising bucket, which misguided many observers.

One thing to note is that email budgets are often small – disproportionally small given email's significant return on investment (ROI). A 2009 study[xi] my agency did along with ExactTarget found that over 40 percent of clients stated they had $100,000 or less of their annual budget dedicated to email marketing.

As we evaluate the change in spend projected through 2016 we should note two things:

---

xi http://www.emailstatcenter.com/Research/EmailStatCenter_CompensationAndResourcesStudy.pdf

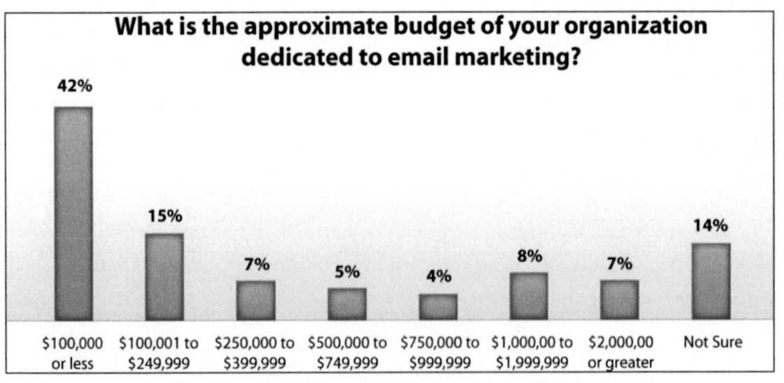

Compensation and Resources Study via BrightWave Marketing, EmailStatCenter.com, and ExactTarget

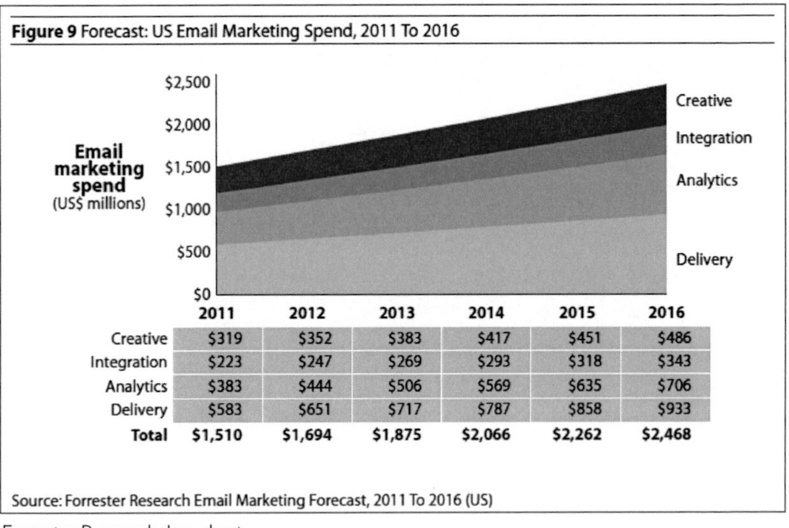

	2011	2012	2013	2014	2015	2016
Creative	$319	$352	$383	$417	$451	$486
Integration	$223	$247	$269	$293	$318	$343
Analytics	$383	$444	$506	$569	$635	$706
Delivery	$583	$651	$717	$787	$858	$933
Total	**$1,510**	**$1,694**	**$1,875**	**$2,066**	**$2,262**	**$2,468**

Source: Forrester Research Email Marketing Forecast, 2011 To 2016 (US)

Forrester Research, Inc. chart

1. Email is certainly not fading away, which many prognosticators have said ad nauseam. It continues to be the healthy heart of digital marketing.

2. The "other" non-delivery areas of email are poised to gain significantly more of the email marketing spend than delivery. While all of the major ESPs have diversified and have largely strong businesses, this represents somewhat of a land grab for the key areas of email marketing services. Agencies (specialized and broad), tech startups, consulting firms, and the ESPs all have

begun to make a play (or are noticeably absent from the space) for these other dollars.

Now we'll take a real look at what is changing on the email investment front and where the smart money is going.

Where Should You Make Your Email Marketing Investments in a Changing Industry?

Let's look at where that money is going and what has changed in the past five to 10 years in terms of email investment areas.

Strategy

Spray-and-pray and batch-and-blast were the de facto email marketing strategies as we headed into the millennium. Contact strategy, strategic blue prints, segmentation, privacy, and preference considerations, as well as dozens of other optimization areas fill the current email road map with a plethora of new opportunities. My firm sees first-hand how some of these optimizations can lead to huge new revenue gains. For one e-commerce client, key changes we made led to over $1 million in additional revenue. You can't find that kind of money if you're not looking.

Creative

Creative often has the dominion of a general creative group or advertising agency. As a result, it historically has been an afterthought, which is why five years ago email creative was a JPEG off of an offline marketing piece and then coded and dropped into an email delivery platform. The nuanced needs on the design, messaging, and coding have called for these practices to end, and they largely have. Not always to the level that they should, but I believe email creative has improved drastically in the past few years and it has also birthed a well-prized employee for any email-centric firm – the email designer.

Production/Campaign Management

Many self-service email delivery platforms continue to do bang-up business while many seem to be hanging on by a thread. Regardless, the business of managing email campaigns is a hot area. Part staff augmentation and part subject matter expertise that can't be found or done in-house is an in-demand area of the email industry. While many brand-side email marketers managing the treacherous and never-ending email campaign cycle have fled to the safer waters of social media, or have been promoted out of campaign management, many of these brands find themselves without the bodies or brains to get revenue-producing emails out the door. Look for more brands to partner with specialized firms and let them handle this dirty work.

Analytics

The perennially underutilized area of data in digital marketing seems to be changing. While many still leave 90 percent of their email metrics unseen, big data and more demanding CMOs are changing this. Email analytics are increasingly more connected to other marketing silos which are making email more accountable and primed for major investment if brands can connect the dots on what is working and what isn't.

Mobile and Social

These two areas are obviously game-changers in the digital space but are attracting special attention and investment as they relate to email as well. Dozens of companies have surfaced that are tied to enhancing email's impact as it correlates to mobile and social (think social data and tools, smartphone rendering, etc.).

Channel/Technology Integration

Whether via an API (application programming interface), point of sale, or other elements of technology integration, this will always fuel investment, but of course nowhere near a typical technology-integration level. Some companies consider integration tied to their email program to be a political issue and some invest heavily, ensuring their email program is tied to many arms of the business.

Acquisition and List Management

This is an interesting one, and I can already envision the representatives of related companies commenting that the days of spending money to hit some list-growth numbers are largely gone. I like this space for what it isn't. It isn't 2002 or even 2009. This bucket now consists of compelling opportunities to grow your list in relevant, measureable, and targeted methods. Lots of new and interesting vendors are worth a test drive in this space, too. Of course, keeping the list clean is essential and requires attention and often budget.

Staff Augmentation

Many agile marketers find it is easier to rent than staff a robust email program. While our industry doesn't have the deepest bench in the world in terms of talent, many brandside marketers know they can house all email program functions in-house and prefer to partner with a specialized firm that can perform some of these highly specific areas as an extension of their team.

Email investment, like most budgeting items, can be highly subjective. However, successful email marketers will focus their investment on actually improving their email programs as opposed to merely ensuring that the send button is pushed. Make sure you ask yourself this question at least once a year: Where are you spending your money and what investment should change in the next five years? You may find yourself making different decisions in terms of investments and partnerships.

---

7 Questions About the State of Email Marketing

I have run my email-focused digital agency for 10 years and I have heard thousands of questions from clients, prospects, peers, and friends about how email marketing really works (and doesn't). This is my second book that hopefully addresses most of these questions (and more). Many of the questions stay the same (best day to send an email, good open rate, etc.), but many new ones are popping up as the industry evolves. Here are eight questions (and one history nugget) that are relevant to today's inbox and are good to have in your back pocket.

1. I have more Twitter followers than email subscribers - does that mean Twitter is more valuable to my brand?

Without a doubt, Twitter is an easier platform to deploy messages from – no creative, coding, delivery issues, and the like. But it depends on your goals and results (you are getting metrics for your Twitter posts, right?). It's a bit like arguing about which is better: coffee vs. lunch – they serve different purposes. However, most consumers won't engage with (much less read) your Twitter message. Brafton reported only 6 percent of tweets sent across the web see even a single retweet, which is for many the key indicator of engagement on Twitter.

2. Will email die if the daily deal suffers the fate of the dot-com bubble?

While the daily deal space (i.e. Groupon, LivingSocial and Scoutmob) has a lot in common with the good old days of 1998 to 2000 (and possibly the unfortunate aftermath of the bubble), consumers are saying they can't get enough of the coupon emails, at least for now. Sixty-one percent of surveyed consumers said they read all of the daily deal messages. And most access the emails at least once a day. Even if 90 percent of the daily deal sites die (and they will), email will be just fine, thank you.

3. When was the first email sent?

The first email[xii] was sent in 1971 between two computers right next to each. It is fair to say that email is a mature digital communication when compared to other digital message mediums.

4. How are email and social subscribers and followers different?

The short answer comes from our friend and digital marketing expert Jay Baer[xiii], president of Convince & Convert: "Facebook 'likes' and Twitter 'follows' are not 'subscriptions' at all, but rather assertions of passion. Social media connectivity is digital bumper stickering." I would add that in these channels, consumers are giving permission for different

---

xii http://en.wikipedia.org/wiki/ARPANET

xiii http://www.convinceandconvert.com/email-marketing-advice/why-social-supports-email-in-the-interactive-marketing-hub/

reasons, so treat them and communicate with them differently based on these very different mindsets.

5. I have an email list that I inherited – permission was not granted to email these people. Can I email them?

Legally, yes, assuming you follow all elements of the CAN-SPAM Act. From the view of "should you?" it's a bit trickier since permission email works when there is...permission. So at a minimum you need to obtain permission. Some may choose to send a one-time opt-in email while others may try to communicate through other channels and drive them to a preference center. It's a slippery slope otherwise if you try to skirt around the permission issue.

6. I just switched email partners. All of a sudden after my first mailing with the new vendor, my metrics changed rapidly – the email was the same we sent with the old vendor, so what gives?

Big problem here. It's shockingly basic but difficult to fix. Email service providers have varying ways of measuring and defining metrics. Even click-through percentage can be counted differently - one may say it is clicks based on number of opens while others may say it is based on the number sent or successfully delivered.

7. How come my emails show up without images for some subscribers and look awesome for others?

According to MarketingSherpa, only 33 percent of consumers have images on by default. So it depends on the email client they use (and the version) and how you design and code your email. Make sure you are comparing apples to apples here.

"Conversations among the members of your marketplace happen whether you like it or not. Good marketing encourages the right sort of conversations."
SETH GODIN

CHAPTER 2

Email in a Social Age

Social media may be email marketing's cooler little brother and get more coverage in the press and general mindshare in the digital world, but both tend to play well together. In fact, any social media manager will tell you they couldn't have a thriving community without email. Email recognizes this and in some cases has the chance to shine a bit more because of the potential for a dramatic one two punch. For many marketers the purpose, strategy and tactics of their email programs have adapted as many brands move to a more social dominated landscape. This section takes a look at what is needed to bring your email program into the social era.

Integrating Social and Email Marketing: Where to Begin

Email marketers have brushed aside the notion of the email channel's imminent death. Instead, we've fully embraced social as the go-to marketing channel (sorry, search), benefiting both email and social channels.

Email and social media are among the top marketing tactics that will continue to benefit from increased spending. Several studies have found that integrating email and social is one of the most important email marketing initiatives.

Savvy marketers are trying to figure out how to build their databases and communicate effectively and intimately, while also driving traffic and business. However, the do's and don'ts of integrating email and social go beyond plugging in a few links on each platform.

According to a StrongMail survey, 27 percent of those surveyed had already implemented an integrated strategy, and another 24 percent had developed a strategy and were researching how to put it in practice. A total of 18 percent of the executives surveyed wanted to add social components into their email efforts but did not know where to begin. Another 10 percent said their CEO's teenage daughter said they were idiots unless they moved all of their marketing dollars into social widget apps and made her CMO (well, maybe the last one didn't happen).

Here are six do's and don'ts for developing an integrated email and social marketing plan.

Do: Get to Know Your Audience

If you are a business-to-business (B2B) marketer, do your subscribers crave Pinterest and Facebook tie ins in their emails? Probably not. What about content from and links to YouTube channels and LinkedIn Groups? Maybe so.

There is no template for how and what to include in your email offerings. So, where should you start? Develop an outline of your audience, including where they are in terms of social networks and what you can offer them via both email and social channels. Remember, it takes more than a button to have an integrated social and email communication strategy.

Do: Follow Your Customers

Does email marketing drive a considerable part of your revenue? And is it the key medium for your customers to learn more about your company? Then you shouldn't turn your back on email, especially in terms of continuing to build your permission lists.

Organic email acquisition will be the biggest casualty of the social marketing phenomenon. Why? Go to just about any one of your favorite sites – whether it's a content or commerce site. Then notice the plethora of "share," "become a fan," "follow us," and other buttons that didn't exist a few years back (but again please remember a button doesn't make for an integrated email and social program).

What if your customers just use social to connect with their friends and have no interest in social as an advertising medium? They just want good old-fashioned email offers and newsletters delivered to their inbox. Good luck finding a preference center or opt-in signup form, link, box, or button. Email may be the cash cow, but it's too often left hiding in the barn for most sites. Signing up for your emails must be easy to find and easy to do. It is a huge and costly mistake to promote social without email acquisition opportunities nearby.

Don't: Be the Cut-and-Paste Social Marketer

As an agency CEO who sees great financial opportunities on both the client and agency side for social media, there are very few things that get me as annoyed as the cut-and-paste social marketer. This occurs when "gurus" have convinced their company or client that they have mastered social marketing because they update all social media in one click. Under this scenario, if you are a fan or a follower of a company on Facebook and Twitter and other platforms, you get the exact same message.

This approach often spills over to email, though the email messages are lengthier. Each medium, in my opinion, is unique and deserves a specialized approach. Conversations, including tone and personality, should vary depending on each channel. Facebook, for our clients, is a more casual and fun conversation platform, while email is a bit more buttoned-up. Generally, the audiences reflect and drive that notion. And the same is true when providing offers and overall value to customers and prospects.

Brands that cut and paste content and offers on all platforms dilute the value of opting in to each experience. If you send me the same offer on Facebook as you do on email, one will be ignored. Therefore, be sure to alter tone and value based on the conversation. It's worth the extra time and effort.

Do: Share Content and Value

Content and value drive sharing; buttons do not. Including a "forward to a friend" button in an email that the user did not consider worth forwarding doesn't mean you've launched a viral campaign. Nor will your email campaign automatically become social if you add a "share" button or Facebook and Twitter logo to the email. Yes, that may buy you some time from the executive who just realized his company needs to have a social strategy, but dig deeper. I have seen too many emails that have huge share/add, Twitter/Facebook buttons above the fold (with no context added to them) and thus distract from the real purpose of the email.

Sharing is powerful; it can exponentially grow your message reach and facilitate engagement. But if your email stinks, why would someone share it? Test placement and wording, and remember to give a reason for why someone should share the message or become a fan of your brand if they already get your emails.

Do: Open a New Door if Another One Closes

During or after someone unsubscribes from your email, offer the person the option to become a fan on Facebook or follow you on Twitter. This doesn't violate CAN-SPAM as long as it doesn't intrude on the actual unsubscribe process or add an extra step. Some people may just be sick of your emails but want to stay dialed in with your other marketing efforts.

Do: Build Lists Strategically

Social networks are underutilized for driving opt-ins of email programs. Reminding your fans and followers that you offer great value via email (assuming it isn't the exact same content as provided via social) is a no-brainer, yet rarely done. It is much more common for marketers to use their emails to build social databases (and to a certain extent, that really is what you're building via Twitter and Facebook). When done correctly, the impact can be significant.

One of our clients – a major consumer brand – used the success and momentum of its email program to grow its Facebook page into one of the largest in existence today. It did so through testing and making clear that a great community was already in place.

That is the key in making a conversion; a click from your email to a Facebook page isn't a score. Getting a subscriber to engage or become a fan is, though it isn't as easy as most think. Facebook is now making it easier for marketers.

Email Is Social

Remember in the "The Matrix"[xiv] when Keanu is trying to figure out how those telekinetic kids are moving objects with their mind, and then one of them gives him the secret? "There is no spoon." Well, that's what I keep thinking every time I hear someone say "Social is killing email (which thankfully dwindles every quarter)." It's just a mindset; it's not fact or even reality in the way they are articulating. We just have to change our way of thinking. I think it's time to "bend the spoon" in the other direction:

Email is social.

I don't mean email is a way to be social or is "just as good as" social, I mean it is social. You could even argue it was the first "social network," or at least the first way to be social on the Internet.

Let's look at how we, as email marketers, have been managing social campaigns for years.

Right now, we are managing a group of users a brand has collected (CRM database). These users are presumably fans of the brand, and some of their most loyal customers. The way email is set up, a message

xiv http://en.wikipedia.org/wiki/The_Matrix

goes out from the brand and hits each individual, and since we don't share the distribution list, individuals don't share with each other (knowingly), although they likely are sharing outside of this list. They may respond to the brand by replying to the message and they may forward the information on to a friend or share it through another platform. Typically, the brand has no understanding of who that new person is (although that is readily available to them through metrics) or where else the information might lead unless the new person opts in or replies to the brand themselves.

Are social medium campaigns all that different?

Social campaigns basically start out the same way. There is still a group of users a brand has collected (Facebook fans/"likes," Twitter followers, etc.) that a message is sent to; the users can still respond and share in similar ways to the email campaign. However, social campaigns have an increased "conversation" component. It's not that conversation isn't happening with email, it's just more likely to happen with social, and the brand has some ability to witness a conversation between other users.

Philosophically, there is no difference. You are managing content that is pushed out to subscribed (opted-in, following, etc.) users and managing responses (once someone responds, you now have a conversation). These content messages can either be scheduled according to a marketing strategy calendar or sent out in response to some immediate need (PR crisis, pending event, sales, etc.). The message's content still needs to be relevant, timely, and match the expectations of the subscriber in either the email channel or a network channel (and we all know each network is different).

Email marketers need to get in this mindset – it's not either/or; it's not one killing the other; it's two different channels that are very much alike.

This is why every facet of the campaign needs to be integrated; marketers need to start looking at holistic messaging strategies. Does your community manager have a schedule of when content will be pushed to Twitter and Facebook so that those two channels are complementing each other? Is email on that same calendar? I'm not just talking about someone, somewhere in your organization who has a master marketing calendar. I mean, is all of your digital messaging working together in real time? Email is just another one of your social channels, and it could be argued that the people who subscribe to that list are your best customers.

Email should be working hand-in-hand with the other digital messaging channels, and each channel should complement and leverage the strength of the other. Ask yourself:

- Are you talking to your best advocates (email list) and pushing them to your Facebook page where they can amplify the conversation in your favor?

- Are you following up on conversations happening on Twitter and sending your followers the more detailed information, or coupon, or content they want?

If email isn't working in tandem with all these other channels, then you are having a fragmented conversation with your users. By bringing all of these social channels together, you can develop a robust marketing plan that not only keeps your customers happy; it also helps them become advocates for your brand.

Why Email Is Like Madonna and Facebook Is Like Lady Gaga

I interviewed Jay Baer, president of Convince & Convert and author of two excellent books The NOW Revolution and Youtility, about emerging areas of digital marketing and how they intersect with email marketing. Jay is a highly regarded social expert who understands email's yin-yang relationship with the medium.

Jay and I talked about the nuances between email and social, why where email and social "live" matters, and, of course, what Madonna and Lady Gaga have to do with this discussion.

Simms Jenkins: You are the rare social media star who actually finds email not only relevant but essential to a strong social media program.

What do you say to the naysayers who claim email is old-fashioned and irrelevant in today's social age?

Jay Baer: I wouldn't call myself a social media star, but rather a social media strategist who happens to have a blog some people read, but thank you. I speak and write about email and its integration with social media because I believe it to be strategically sound, and because of my long background (1994) in digital and email marketing. Is email as popular among young people as other forms of communication? No. But it's still the most viable, measurable, reliable way to communicate to people who have asked you to do so, and that's not going out of style any time soon. As I've heard said, it's pretty tough to say email is dying when you have to have an email address to even join a social network.

Simms Jenkins: You previously told me that email is Madonna and Facebook is Lady Gaga. Can you expand on that?

Jay Baer: Madonna is the original. Gaga is the newfangled upstart. But fundamentally, they are the same thing with a fresh coat of paint. Email and Facebook have the same dynamic. Both are used to keep your business top-of-mind among people who have given you permission to do so. Realize that 84 percent of Facebook fans are current or former customers of a company (DDB, 2011). Thus, you are preaching to the choir on Facebook (most of the time) – just like you are with email.

Simms Jenkins: What does social do that email can't and vice versa?

Jay Baer: Social certainly has the advantage of real-time interaction, and the ability to create back-and-forth conversations where the "fan" (subscriber in the world of email) can create content too. Email has the advantage of being more reliable. Your open rate (a dicey metric anyway) might be just 25 percent, but something like 99 percent of your subscribers will actually receive the email. On Facebook, the percentage of "fans" who see any particular status update from you is 7-15 percent (depending upon whose data you believe). On Twitter, who knows? But, if you think that 53,000 people see every tweet I send, you're delusional.

Simms Jenkins: So will email and social continue to be siloed in corporations despite the obvious similarities?

Jay Baer: I hope not. But the reality is that in larger companies, email is often still a child of technology or direct marketing, whereby social is more in the communications camp. I see better hopes for synergy among mid-sized and smaller companies.

Simms Jenkins: We know that email drives social media and often is the number one reason people are visiting fan pages and returning to social networks. However, social doesn't seem to be leveraged to assist marketers' email programs, which usually have a higher ROI – why is that?

Jay Baer: It's because people who work in social media professionally often have little experience with email, or interest in growing the email channel. The first thing most companies should do (after adding social media icons to email templates) is to set up an email sign-up form on Facebook. Alas, it's pretty rare to see it.

Simms Jenkins: Great point. Do you agree with the findings that the most active social media users are probably your best email subscribers?

Jay Baer: Of course. The people who like your company for real also like your company enough to become a Facebook fan or subscribe to the email newsletter.

Simms Jenkins: Who are a few brands that are doing some great things in the social and email space in terms of integration?

Jay Baer: California Tortilla ties the channels together very well at the content level. Lululemon does some interesting integration, too. Scotts (the guys behind Miracle-Gro) do a great job, too.

Simms Jenkins: What about the email purists who deride social's value as just noise and fluffy engagement? Is there validity to that statement or are they just feeling threatened?

Jay Baer: Email professionals who willfully ignore the 847 million members on Facebook (at publication over 1 billion), and the 500 million Twitter accounts are fooling themselves. Put it this way: the average Facebook user spends between six and seven hours per month on it, and with pleasure. People check their email a few times a day, but begrudgingly. Social networks change and/or improve their features and user experience every single month. When was the last time email got better in any meaningful way? That's the problem email faces in the future...it never got any better.

Simms Jenkins: What is your advice to a digital marketing professional who is confused by the tug-of-war game often played by email and social media?

Jay Baer: The goal isn't to be good at social media. The goal is to be good at business because of social media, and integrating with email is a great step in bridging that gap.

Simms Jenkins: Thanks Jay. It's always a pleasure speaking with you, but I appreciate your clarity on this important and often divisive topic. I hope each and every digital marketing practitioner can learn from this and figure how to better seamlessly tie all of their digital programs together, not even just email and social. Let the silos fall down!

Social Networks and Bacn – The Good, Bad, and Ugly

As social networks increase their use of email marketing in order to drive more site traffic and user frequency, bacn[xv] has come back into the headlines. Bacn was described by The New York Times back in 2007 as "Impersonal e-mail messages that are nearly as annoying as spam but that you have chosen to receive." Bacn is one example that proves that email has not been murdered by social media but instead is sizzling hot due to social networks' increasingly strong reliance on email marketing to get its users back and active on their sites. So blame social media for the extra order of bacn during the last few years. Even Dow Jones MarketWatch interviewed me for a radio segment[xvi] on the surge of bacn.

These types of emails are often dismissed as annoying, yet they drive traffic, high response rates, and ROI, and for the end subscriber (oh yeah, us email marketers often forget about her) they provide something

---

xv http://www.clickz.com/clickz/column/2044487/steps-cook-bacn

xvi http://www.marketwatch.com/story/bacon-clogs-e-mail-arteries-too-2011-06-27

essential to the relationship you have with the sender. Sometimes it's a shipping confirmation, a welcome email, or statement balance. With social networks as the hub for many Internet users, email is the bridge informing you of everything under the sun as it relates to your friends, followers, and connections, all with the goal of getting you back on their site.

The bacn messages most certainly take up space in your inbox, but you want them and gave permission, right? Let's take a look at how the three big social networks bring home the bacn.

Facebook

Facebook has over one billion users and almost 700 million mobile users that check the site everywhere and often. Besides the addictive nature of the social network, what gets its users to come back? Emails notifying that a user commented on a photo, a new friend request, or your ex-girlfriend's birthday. That's bacn.

Twitter

Twitter has followed this trend and raised it once playing up to the average tweeter's sense of vanity and needing to know what is happening in real time. Emails are now instantly sent notifying a user when they are mentioned, retweeted, and followed. This is some self-satisfying bacn.

LinkedIn

The best email marketer of the social trinity, LinkedIn is constantly adding new features to its email mix and it's geared toward empowering the user with choice and information (overview of your network's activities).

LinkedIn also uses the Amazon styled methodology of "you bought this, so you may be interested in that" in its connection confirmation emails, suggesting other new contacts from your new connection's network as well as companies to follow (a relatively underused feature).

The popularity of these social emails has resulted in phishing scams that look like a real piece of bacn from these networks which are dangerous since so many consumers won't pause for a moment to click on the email that looks like it's from Facebook. So keep an eye out for that and make the most of this potentially satisfying email dish.

Tips and Best Practices to Follow if You Deliver the Bacn

- Provide users with the ability to control frequency, type of messages, and other settings so these don't backfire and send the subscriber running to your competitor.

- Leverage the treasure trove of data to customize the message to increase the value and relevancy of each and every message.

- Ensure you have designed, coded, and tested your messages to still deliver the essential information when a subscriber has images turned off.

- Make sure your subject line clearly teases the purpose and value of the email.

- Have your marketing people own these, not IT, and remember to update creative and cross-promotional elements frequently.

- Don't have automated messages that are essential to someone's relationship with your company segue into more promotional emails without getting a true opt-in. I would argue that LinkedIn Today crosses a line since I didn't sign up for it and I clearly don't "need" it for my LinkedIn membership. CAN-SPAM then becomes applicable as well, since it's not considered a transactional email and not fully CAN-SPAM compliant.

Inside Your Marketing Department: Who Is the Rightful Owner of Social?

Let's skip the obvious buzzwords and generalities. We know social media needs to be engaging and authentic.

List quality is more important than list size; and content, offers, and relevancy trump all.

Of course, your customers own social, but who is the rightful channel owner within your organization? The likely suspects are generally public relations or the digital team and, more specifically, your email group. While agencies, consultants, "ninjas," and the like are playing an important role in this nascent channel's management and strategic evolution, someone within your company needs to own it, right?

While the public relations side of the house was slow to figure out their place at the table during the first digital boom (although press releases did have an infamous role in the dot-com boom), in recent years they have moved quickly to say, "Social is all about content and public relations is about content, so hey, sounds like a good fit." And good for them.

Digital teams, whether generalists or specialized forces of search, mobile, email, and web, have been much more in the spotlight during the past decade or so. Their budgets and internal importance have likely grown in a significant manner.

While there is no absolute rule for every organization, who do you think should "own" social media within a typical marketing department?

Well, the early evidence and smart money seems to certainly be on digital, with the email program folks the likeliest and most appropriate owners. So why?

- Email and social are likely the only two, if not most important, permission-driven marketing programs for any company. While reach is important, permission means qualified communications and conversations.

- Email should be about engagement and revenue. Social should deliver these.

- Email should be about targeted and measurable messaging. Social should follow in these tracks.

- The best and most integrated digital programs have email and social as the spark and fire. Therefore, they need to collaborate tightly and serve each other.

- Email and social programs and related campaigns and conversations don't ever stop, and in order to maximize the quick-moving nature of these mediums, you must have a team that can be nimble, versatile, and detail-oriented. If you are familiar with the life of an email rock star[xvii], this will make sense as the social process flow is similar to email and unique when compared to other marketing disciplines.

Some organizations have cutting-edge outreach, advocacy, and PR departments that do more than generate press releases, so they should certainly make their case for playing a key role in their social media program. However, for the day-to-day ownership and management, I think email's gatekeepers have proven to be the rightful heirs to the social thrones. Of course, this all changes when you mismanage it or go to market without a strong strategic plan with real business goals. That's the challenge for anyone remotely leveraging social media.

In both email and social, it's all about your end user, not you! This is often hard for marketers with a plan and a goal to remember as they bombard timelines and inboxes. Respect the customer and deliver value on any platform and you are two steps ahead of most.

Let's go back to my opening paragraph and see how these social media essentials match up to email marketing essentials:

- Social media needs to be engaging and authentic. A must for email.

- List quality is more important than list size. Very true for email.

- Content, offers, and relevancy trump all. Any email campaign metrics will validate that this is the secret sauce of email marketing.

---

xvii http://www.clickz.com/clickz/column/2040772/appreciation-email-rock-stars

"Mobile is a lot closer to TV than it is to desktop."
MARK ZUCKERBERG

CHAPTER 3

Mobile's Impact on Email Marketing

I believe nothing will impact email more profoundly and more positively than the growth of mobile devices. This has already changed how, where and when consumers and businesses interact with each other, and email is a major beneficiary. Zuckerberg, the Facebook founder, and his quote above demonstrate that consumers will change their habits in a major way and marketers will follow. This section will help guide you through the process of "mobilizing" your email marketing program to ensure you will be able to follow your customers wherever they may lead you.

Being a Mobile Marketer Who Happens to Be Leveraging the Email Channel

It's what mobile consumers are doing, and they are the future of digital. According to Pew, email is the most popular activity on smartphones and tablets. Yep, not Facebook or search. The good old-fashioned warhorse of digital communications. Of course, some want to make this warhorse turn into a unicorn with mobile magic.

Email marketers *are* mobile marketers whether you like it or not (go ahead and add it to your LinkedIn profile right now). I often hear that our audience isn't reading on smartphones. Knotice says the number of emails opened on a mobile device (smartphone and/or tablet) during the first half of 2012 overall rose to 36 percent. My agency finds the number around 50 percent for some of our clients. Either way, it's increasing for everyone. If you are in the dark on your audience and their mobile readership, make that goal number one.

Changing your mindset would be goal number two. Mobile is the new reality and should greatly impact all of your email marketing efforts going forward.

New tricks, technology, and savvy testing can accomplish a lot on the mobile front. In addition to understanding your audience and building a game plan with that in mind, the execution of your campaigns to a mobile readership is crucial.

The right message on the right device can be the difference between a read, a click, or a purchase. Leveraging our agency's proprietary technology, check out our holiday card, which served up multiple versions based on how and where you were reading our email.

Desktop Version:

Mobile Version (next page):

While the difference between a vertical and horizontal email may not seem like much to say, an e-commerce marketer, the user experience is what makes the difference. Your merchandising strategy is going to differ in a small, suburban mall versus a large, pedestrian-friendly urban location. Shouldn't this also be the case for email messages sent to smartphone and desktop subscribers? Think about how you can place products and calls to action in a different manner on an email targeted to an IOS device versus a more traditional templated approach in a desktop version.

A smart email marketer is going to explore what moves the needle on timing and segmentation of each campaign. Mobile email opens up a whole new can of worms – we have found response times increase greatly at certain times of day for mobile users. A sound segmentation strategy comes out of this kind of knowledge so you can consider segmenting groups based on behavior or device.

The possibilities are plentiful and strong as email marketers need to embrace this new way of thinking to become the best mobile and email marketer they can be.

The 7 Golden Rules of Mobile Email

1. *Smartphone owners are more likely to read emails than make calls.* Smartphones are largely email computing devices for many busy consumers and professionals. In fact, the 2013 Digital Publisher Report from Adobe reported that among 18-54 year olds, 79% of smartphone owners used their device for email more than for making calls. It should also be noted that email is the number one activity for tablet owners as well.

   So what's the takeaway here? Email is only getting more important, and your email program better be adapting to this new future. Pretty soon, 50% or more of most your email subscribers will be reading your email on their smartphone. Time to get mobilized.

2. *Consumers read more emails on mobile devices than a browser or desktop.* Most data is showing continued growth in how emails

are being read – primarily mobile as opposed to the more "traditional" methods of desktop and webmail. This means your emails look different to many consumers which leads us to #3.

3. *How your email looks on smartphones matters to consumers.* Blue Hornet's 2013 study showed 80% of consumers delete a mobile email when it doesn't look good and 30% unsubscribe, up from 18% in 2012. That means almost 9 out of every 10 subscribers are having some kind of negative action based on an appearance. In the email world, unsubscribes and deletes are about as negative as you can go. Combine this with previous research from Merkle and Blue Hornet that ties bad emails to negative perceptions of the brand and most marketers should have serious ammunition to invest in their email to ensure it doesn't look "bad" on mobile devices.

4. *Understanding where your subscribers read your email to determine the right approach.* This might seem like step #1 but I believe you must understand the macro picture before dialing into the correct approach for your brand. You should be able to get subscriber logs from your email partner or vendors like Litmus or ReturnPath. Understanding how many subscribers are reading your emails on smartphones and tablets is equally as important as understanding what operating systems your audience is on. Understanding how your subscribers read your email will let you define your next steps.

5. *What you are trying to accomplish in your mobile email impacts your approach.* Some great mobile emails that build awareness would be horrible communications when the goal is to drive sales. Building these emails must be driven by your business goal, not any best practices you may have read about (even here). Speaking of mobile commerce, every marketer must be aware of this metric from Adobe: "Mobile purchasing decisions are most influenced by Emails from companies (71%) only surpassed by the influence of Friends (87%)."

6. *Device matters & behavior varies.* Responsive design, where media queries help an email (or web page) adapt its layout to its

viewing environment (which in the mobile email world is predominantly iOS or Android), does some nice things but it doesn't help deliver the right email to the right device the way some new technology can. My firm, as well as Movable Ink, has developed unique technology that allows a brand to serve up multiple versions of an email and have the "right" version display on your mobile device. This is a possible game-changer within a game-changer.

7. *Another take-away tied to #4 is don't just bet on the iPhone/iOS.* Do your homework on your own database, and ignore Droid subscribers at your own risk. To illustrate this point, the talented creative folks at BrightWave Marketing developed an infographic to shed more light on this subject.

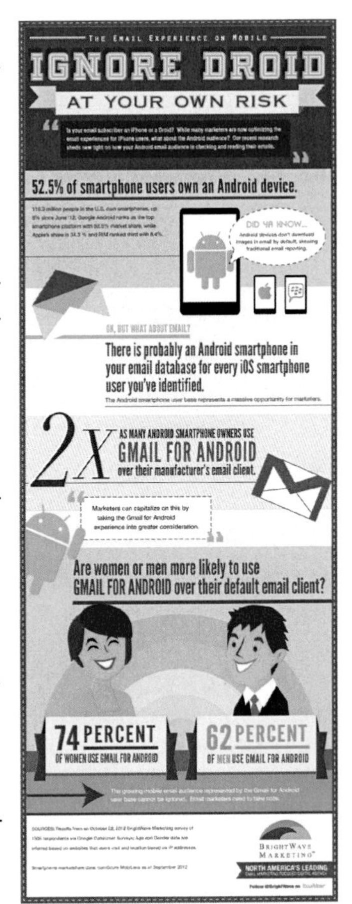

Mobile's Pending Revolution of the Inbox

For a fresh perspective, deep passion and real know-how, I talked with two of the nation's foremost experts about mobile's impact on email marketing.

Justine Jordan, marketing director at Litmus, and Jay Jhun, vice president of strategic services at BrightWave Marketing discussed how email marketers need to adapt to mobile and provide insight on how savvy marketers can make the leap and embrace subscribers' changing email consumption habits.

Simms Jenkins: Why should a digital marketer care about mobile email as opposed to sexier mobile topics like apps, display, and the like?

Justine Jordan: Maybe they should, maybe they shouldn't! Like so many things in digital marketing, deciding to spend time and resources on mobile email should factor in your goals, product, service, and audience preferences. Ignore media hype and look at your metrics. I'm a big fan of making decisions based on data (although a few people have gotten lucky on hunches, I'm not yet one of them).

Look at your web analytics, email analytics, and any other dashboards at your disposal to make a smart decision around where to invest. If email makes up a significant chunk of your revenue or you already have a mobile app, chances are that there's an opportunity in taking a closer look at making your emails mobile-friendly. Design is sexy, and creating beautiful and usable email experiences can pay dividends over more intensive projects.

Jay Jhun: As more and more people engage with digital communications and media via smartphones, marketing decision-makers need to at least start planning to invest in mobile email templates, landing pages, and websites and including mobile interactions and conversions into their campaign plans. Nobody is technically "late to the dance" when it comes to mobile-optimized emails but, to Justine's point, now is the time to be studying your customers' mobile behavior through web and email analytics. Your email program is one of the best places to begin your mobile discovery agenda because of the ability to strategically tag and test content.

Simms Jenkins: Well said. Do you think the trend that email consumption is making up about almost half of every hour of smartphone usage will continue with email being the dominant thing consumers are doing on these devices? Or shrink? Why?

Justine Jordan: Once again, I take these surveys and stats with a grain of salt. What is your audience doing for every hour they spend on a smartphone? Instead of speculating, find out. If they are spending half of that hour in your app, then do you care about what they are spending the other half doing?

That said, I don't expect email usage on smartphones to shrink any time soon.

Jay Jhun: If there's any competition to be had in where people spend time on mobile devices, it looks like it should come from social media, photography, and game apps. That being said, as more mobile web experiences come online to support mobile email interactions, there's no question in my mind that email apps will carry a big percentage of that hour.

Simms Jenkins: What are the top five recommendations you would provide to anyone looking to ensure their email program adapts to the mobile world?

Justine Jordan:

1. Learn how your subscribers/customers are interacting with your brand in a mobile world. Are they visiting your site? Trying to buy from you? Reading your emails? Let this data be your guide.

2. Modify designs to be finger-friendly and readable on small screens. Even your existing content can be made more presentable and usable by enlarging fonts, introducing whitespace, and increasing contrast.

3. Simplify and prioritize. Cut down extraneous content, links, navigation, and information.

4. If you're going to optimize your emails for mobile devices, here are a couple things you shouldn't do:

    - Forget about your landing pages. The worst thing about finding a gorgeous mobile-friendly email in my inbox is clicking through to an unusable website or landing page.

    - Segment by mobile device. Instead of creating device-specific versions of your email, send one email that works well in all viewing environments and devices. An exception to this rule is if you're sending emails about OS-specific mobile apps.

5. Read "Mobile First" by Luke Wroblewski. It might change your life.

Jay Jhun: Another thing I love about the emerging mobile age is that it is bringing fresh attention to the same questions about best practices in email campaign design and execution. More marketers could do better in these areas. Fact of the matter is that the "age old" fundamentals hold true, regardless of environment:

1. Subject lines are really important.

2. Your primary call-to-action should be a button.

3. Deliver value.

4. Talk to your audience as if you know something about them (i.e., be relevant).

5. Test and learn.

Simms Jenkins: Seventy percent of consumers will delete your email if it doesn't look good on mobile devices, according to BlueHornet. That is a significant number, yet most emails don't look anywhere close to good on smartphones. Why?

Justine Jordan: Consumers might move on and come back to read the email later when their viewing environment is better suited to the email. But if this is true, it means that consumers are becoming less tolerant of tiny fonts, tons of copy, and emails that simply can't be read on a mobile device. Consumers don't want to work to read your message, so make it easy for them!

Jay Jhun: Perhaps what we as marketers really want to understand is the positive version of the posed question: Would they be more likely to take action (like an online purchase) on a promotional email that they receive on their mobile device if it were optimized for viewing and usability? iPhones are an easier nut to crack because they're easy to detect and content is scaled to fit the screen. The plethora of Android devices and operating system versions (Ice Cream Sandwich, Honeycomb, etc.)

begs the question of what the most common email experience is outside of the iOS platform. (At the time of publication, ABI Research projected over 1.4 Billion and 268 million active smartphone and tablet users respectively. Android is estimated to have a market share of 57% and Apple's iOS to have 21%) .

Simms Jenkins: Can you talk about expected results when you design and code compelling mobile emails?

Justine Jordan: Honestly, hard case studies are just starting to come to light. It seems like 2011 was the year of discovery when it came to mobile email. 2012 was the year of optimization, action, and results! I've heard some promising anecdotes of how mobile-friendly emails have increased click-through and conversion, especially for retailers!

Jay Jhun: My expectation is that click rates would increase, in general, when emails are optimized for mobile. Open rates are driven by things like subject lines, frequency, relevancy of the subject line, and whether images render or not. When comparing conversion rates for desktop vs. mobile devices, I would expect to see varying degrees of improvement for mobile emails skewing more positively when mobile-optimized destinations (websites, videos, etc.) are present.

Simms Jenkins: What evolution might we see on the mobile email front?

Justine Jordan: There's a lot of opportunity here for technology-friendly brands to incorporate real-time data and technologies to communicate with consumers who are "on the go." Combining triggered emails with location-based technologies could result in some pretty cool campaigns. I'm also hoping we'll start to see more brands take advantage of mobile opportunities by incorporating games, apps, video, and other interactive elements into email, since most mobile devices are much more capable than their desktop counterparts when it comes to email rendering.

Jay Jhun: One limiting factor to how email consumption might evolve is the content that marketers have available to them. For example, video content could play a huge part in driving mobile email engagement, but there's a shortage of content. Imagine a restaurant's head chef introducing the latest healthy menu items via video – it would be much more engaging than just delivering the benefits in writing.

Another area where I expect marketers to do a better job over the next couple of years is in driving clicks from emails into mobile apps. If I get an email from Amazon about something on my Amazon Wish List, I'd much rather complete my purchase on my device than on a mobile website because my phone is still more secure and easier to use.

Simms Jenkins: Parting thoughts to marketers as they enter this new frontier of true mobile email marketing?

Justine Jordan: I think the most important thing for marketers/brands to realize is that "mobile" and "email" aren't things that will be going away any time soon. Email isn't dying, and "mobile" isn't limited to just smartphones. Technology and culture are rapidly evolving, and it reaches out and affects us as human beings – it even changes our behavior. Even if you thought you knew your audience, chances are they are changing right before your eyes as they learn to adapt to new ways of life and new products coming to market. Just as soon as you optimize your email marketing to be more readable on a smartphone, the next big thing will come along and you'll have to figure out how to deal with that, too. Be flexible, be scalable, and don't be afraid to experiment while staying true to your brand.

Jay Jhun: My charge to marketers would be to do the due diligence now in discovering the size of your mobile audience across all channels. Determine the opportunity cost of not investing in mobile optimization and then you'll be able to make the case for budgets that will help evolve your email program for the mobile age.

"The purpose of a business is to create and keep customers."
THEODORE LEVITT

CHAPTER 4

Growing Your Email List

What good is an amazing email program with a tiny list? Every marketer wants size and scale and to really make an impact you need a relatively decent sized list. While I believe in quality over quantity, there are many low hanging fruits for marketers to use to increase the size of their email lists. This chapter will focus on specific ways you can grow your database.

Follow This: How to Create a Compelling Reason to Gain Permission

Let's face it – every web page, application, or brand is screaming at us to like them, follow, check in, connect, sign up, or register. It can be daunting.

Here are eight key questions to ask yourself (or questions your customers will ask you) to make sure wherever your customers and prospects are interacting with you, it's easy for them to do whatever they came to do. Chances are if you accomplish that, your business objectives will have a better chance of occurring.

Questions the marketer should ask himself:

1. What's your goal?

As baseball legend and life philosopher Yogi Berra said "If you don't know where you're going, you might not get there." Goal setting is mission critical for email marketing program success, yet sometimes gets ignored.

So you have to be clear with your vision for any opt-in marketing program. What are you trying to accomplish? How will you do it? How does this impact your business goals? Is it realistic?

Answers will vary for a mature email program versus testing a Twitter proof-of-concept campaign, but you need to ask these questions no matter what. Why? Because someone will eventually.

2. What's the situation?

Just a button, tab, or neat font may be a start, but you must make it easy and clear for your visitors. Always make the opt-in (be it email, Facebook, Twitter, or Foursquare) highly prominent on your site. Southwest Airlines does a nice job in grabbing your attention and then hopefully converting you after the initial click.

Don't waste great opportunities like confirmation pages, welcome emails, social conversations, and more to cross-pollinate complementary permission-based marketing offerings. For example, does your welcome email provide links to your social media presence? Does your Twitter page offer clear direction on how else to engage with your brand? After you buy or opt in, it is very appropriate to want to delve in deeper, and these opportunities are missed by most for leveraging integrated acquisition efforts.

3. How will I measure the success and value?

In addition to "Is my open rate good?" and "When should I send my emails?" we often hear "Is my email database big enough?" Yes, in the numbers-driven world of marketing, size does matter. But how about measuring how many fans or subscribers interacted (measured by likes, comments, or clicks) with the brand in a particular month? What about measuring churn across the various opt-in platforms? What about ROI on each campaign or channel? Content and value of your communications, for better or worse, is often the driver of these metrics, so you want to be sure you are evaluating why and what may have helped increase engagement and sales or decreased call center inquiries.

Questions your customer will ask you:

4. What will you deliver?

Just asking for an email or a like isn't enough in today's marketplace. You must be clear to the end user on what they will receive in exchange for their permission. Regardless of the platform, any company needs to spell out what you will be receiving. One thing to note: marketers may be enamored with social's potential for going beyond offers and discounts, but that's not what's driving them to like you. Don't worry – it's them, not you.

5. Why should I?

Offering incentives to sign up or like a brand is a proven acquisition method. Making them consistent with the brand experience and not bringing in the "wrong" type of user is key too. Regardless of an incentive, you should articulate how the potential subscriber, fan, or follower will benefit from opting in. Is it exclusive content, discounts, or updates from the CEO? Be clear and tell them out of the gate.

6. Do I trust you?

Privacy policies are one of those things it is good to have even though no one reads them. But what about being frank and making it clear about how you will (or won't) use any data from the relationship? Trust is often gained through simple and transparent wording, not legalese or marketing speak that begs the question what are they hiding?

7. OK – you had me at hello, now what?

Everyone hates to fill out long forms, so why don't email marketers learn from the ease of one-click opt-ins like those Twitter and Facebook offer? Capture the email address and then provide the ability for the user to offer more information in order to build a profile. Forcing them down an eight-field form is often one of the biggest reasons email programs don't grow.

8. Is this all you have to offer?

I love to think that any permission-based platform, whether it is Twitter, Facebook, or email, is really a unique channel that offers VIP benefits that can't be accessed elsewhere. The reality of the situation is very different. Your best customers crave something unique, so give it to them. It doesn't have to be monetary offers. Maybe it's a heads-up on a special that will launch in 24 hours or a special survey so your customers can comment on specific topics, and provide compliments (gasp) and their meaningful feedback. Just don't be the cut-and-paste marketer.

Whatever your opt-in marketing platform of choice, remember to stand out, speak clearly, and be respectful to your audience.

Email Acquisition Tactics – Fair Game or Dirty Business?

Nothing gets the email marketing crowd going like the topic of adding people to brands' email lists without their permission. Rightly so, since most significant strength of email marketing is that permission is provided by the subscriber to the brand to send them advertising! It's also what makes it wildly successful as a marketing medium. What it loses in reach, it makes up for in targeting and measurability when compared to its brethren.

Many marketers think a name is a name and a list should be bigger rather than smaller. Other folks say that if you have some kind of interaction with someone in some form (I know, intentionally vague), sending them email communications is fair game. In the B2B world this can be the standard. But should it be regardless of who your company is marketing to?

Let's look at a few real-life examples.

I have attended a few conferences recently and at one I apparently invited several brands into my inbox. However, that was neither the case nor my intention. In fact, I don't even remember having any kind of interaction (such as entering a contest or dropping off my business card in a fishbowl). I know many media companies that host trade shows will sell or include in a sponsorship the names of all attendees. It is somewhat tolerable to get some random emails from these brands right before or after an event, assuming they have some kind of value and/or connection to the event.

However, a few have jumped the shark and I don't condone that. A few just flat out opted me in to their email program without that kind of interaction that is considered at least somewhat in the grey area. For example, Herman Miller put me in its welcome series (or I assume so by its content). The preheader is a bossy declaration: "Look forward to receiving the latest information, incentives and inspiration for your office space." Well, thanks for asking.

If a brand was going to employ this tactic, at the very least it should include some kind of explanation such as, "You are receiving this email as part of your ___ Conference benefits and we hope you will find our emails valuable." Or at least some kind of front-and-center opt-out. Just dumping me in the database doesn't fly without an explanation, even though I can read between the lines. And this doesn't even touch on the creative for this email.

I've also noticed a trend with some especially crafty people using LinkedIn as an email database facilitator. After connecting with some new connections, I have found I end up on their email list. It seems to be mostly small sole proprietor-types versus marketing managers at Fortune 500 companies. Nevertheless, I didn't give them permission to email me about their company when we "connected" via LinkedIn.

What do consumers think of this practice in general? Considering that BlueHornet and Forrester found that 75 percent of consumers did not think it was acceptable to receive emails after they actually *bought* something from a company, I can go out on a limb and think that number would be even higher if the question was, "Is it OK for a company to contact you, that you may or may not know and don't have a specific business relationship with?"

But B2B rules are different, or so many will cry. But should they be?

Shouldn't we generally have the practice of not opting people into a database at all without some kind of permission? Shouldn't this be the standard for any kind of email marketing? At the very least inviting them to opt in rather than just the dump-and-send?

Lost – Email Sign-Ups

What does the epic television series "Lost" have to do with the email marketing world? Trying to connect the two, I started down the path of the smoke monster and spam but like many of this show's plots, I went down another path. Like the magical powers the island possesses, the emergence of social networks has also done wonders for email as the primary traffic driver, but caused the once ubiquitous email sign-up form to disappear much like the island that Jack, Kate, Sawyer, and Hurley have called home for the past few years.

"Sign up for our email newsletter" used to be fairly standard somewhere on the home page for most companies. When you think about it, what site doesn't have giant Facebook and Twitter buttons plastered over the corner real estate where email sign-ups used to be? One would now think that most brands have abandoned email for social as their go-to digital channel.

Of course, email is stronger than ever. Conversely, email acquisition is as weak as ever. Social marketing fever may be partially to blame, but how many of you know where your email sign-up links are on your site? Exactly.

While I'm not an information architect, I do know most executives envision their website as a place for customers and prospects to interact with their brand and get relevant information. Usually, media is driving people here and email serves as an extension of that media buy through the beauty of the opt-in. So, if it costs $4 to get someone to your home page, don't you want to keep talking to them after they leave? The answer is yes, but most marketers are still making that a hard thing to do, even if the site visitor really wants to. Finding an email sign-up can often be a major task and that's a shame.

I looked at the home pages of the top 10 retail sites[xviii] and a handful of other top sites (ranging from Weather.com to The New York Times) to see if they offered an email sign-up on the home page. My takeaway is that most companies are missing out on making it easy and obvious for site visitors to sign up and get email updates and offers.

Of the 10 major sites I visited (including ABC's "Lost" page [xix]), only two had email sign-up offerings on their home page. Three made their social presence available on the home page without providing an email option, and five did not offer either an email or a social call to action. Southwest Airlines deserves special recognition for actually giving email prime real estate twice on its home page and positioning its social options at the bottom of the page (obviously, ROI is influencing site placement at the quirky yet dominating airline and kudos for that happening). In addition, they likely realize that an active social discussion can often be a curse when related to a negative travel experience.

Retailers, who generally depend on email to generate substantial revenue, fared better when it came to merchandising their email program.

All but three of the top 10 retailers made opting in for email an option if you visited their home page. Interestingly, two of the three that don't offer a clear email sign-up path on their home page embody progressive retailing: Amazon.com and Best Buy. An old-school anchor tenant, Macy's is the third. (It should be noted that I have previously discussed this issue with Best Buy marketing representatives - ironically, on Twitter.)

Many marketing executives face a paradox because of the continued emphasis on email list size as a key performance indicator. Often an email program is judged by size alone. While I believe that's a big mistake, you can't grow your subscriber list if your potential subscribers

xviii http://www.experian.com/hitwise/online-trends.html

xix http://abc.go.com/shows/lost

can't find your sign-up forms or if you fail to make it clear as to why they should sign up and what they may receive in return for their permission.

Eliminating email registration from your home page in favor of social networks may please the internal team charged with creating a social presence, but your CFO may be the one to eventually question that. Marketing legend Stan Rapp has said[xx] the value of a single opt-in address is estimated at $118 and Epsilon recently valued an email address at $23 over its lifetime (around four years). Yes, these numbers vary widely, but the point is they are worth something to each business. You can't collect money if you hide the cash register.

Compare this to a study[xxi] by Vitrue, which valued Facebook fans at an average of $3.60 each. That means Starbucks' 6.5 million fan base is worth a bit over $23 million and, using Rapp's email value, Coke Rewards' 12 million reward email subscribers are worth $1.4 billion.

I'm not suggesting eliminating email or social engagement for the benefit of the other. The best email programs are well integrated with social. But if you want to grow your database, make the email sign up easy to find and provide clear value. It is shocking to hear smart marketers complain that their email database is stagnant when it takes four minutes and some good detective work to find a link to an email sign-up page or communication preference center.

Consider a clear email sign-up your magic magnetic property, like the ones that brought Oceanic Flight 815 to the mysterious island in the Pacific. So the moral of this story, like "Lost," may be tough to figure out, but your email destiny can be influenced by the choices you make. Make it easy for them to find you and then you can continue the conversation, and regardless of your math, that is worth something.

---

8 Missed Opportunities for Acquiring Email Subscribers

It's amazing how many leading websites make it difficult to find and/or sign up for their email offerings. I want to continue the theme of

xx http://www.imediaconnection.com/content/22746.asp
xxi https://blogs.oracle.com/socialspotlight/

missed opportunities and now bring attention to the fact that many companies fail to leverage other wonderful opportunities to capture email addresses.

Any company that communicates with customers and prospects outside of their website should be utilizing that customer touch point to potentially acquire email subscribers and other opt-in opportunities.

So, where do you start if increasing your email database and decreasing the costs of other marketing channels are priorities?

Let's first consider eight areas: point of sale, call center, in-store experience, sales calls/marketing collateral, direct mail, mobile, experiential marketing and catalogs and packaging promotional inserts.

1. *Point of Sale*

You may ask, "Don't you want a customer's email before the sale, not after?" If you want to build loyalty, retention, and ongoing relationships and sales with your current customer base, then you absolutely want their email address. Of course, you want them to opt in to future emails so you can send them more information.

Capturing email in a way that doesn't interfere with the transaction and is recorded in an accurate and timely fashion can be a challenge. Methods vary from hand-written notepads (Tuesday Morning does this) to tear-off coupon-oriented slips (Office Depot and CVS employ these tactics) to employees entering your email address into the POS system (like J. Crew does).

2. *Call Center*

The backbone of many CRM programs is usually email marketing or call centers, yet they're rarely integrated from a data collection viewpoint. We all know those incoming customer service calls are expensive. Most companies using call centers are missing out on an email acquisition opportunity. Word to the wise – train your agents on why email is important and list possible hygiene roadblocks prior to using them to assist with the email opt-in process.

3. *In-Store Experience*

Do you place a fishbowl in your restaurant (where it's questionable on whether an opt-in occurs when you drop a business card in a bowl touting a free meal) or do you provide a clear and direct opt-in method for them to be added to your list? We process thousands of in-store opt-in cards each month for our clients and the reason they're effective is they're offered up during the dining experience as a communication option and incentive, provide user-friendly fields and spacing, as well as a clear value proposition on why they should opt in. Finally, these addresses are added in a short time period and customers receive a welcome email before they forget they've opted in. All in all, it's a positive user experience. Therefore, ensure your users have a strong first impression of your email program and don't have to wait for months before getting that coupon or first email promotion.

Be sure to think of every place to potentially take that relationship leap with your customers. Opt-in promotional exposure can be tied to receipts (list a website URL or text to enter short code), displays in retail stores, table tents, menus in restaurants, and even stores that have kiosks or computer terminals.

4. *Sales Calls/Marketing Collateral*

Just so you B2B folks don't think I'm ignoring you, you certainly can leverage face-to-face experiences by driving opt-ins so you can strengthen that long-term relationship. Whether the goal is gaining an opt-in to the corporate newsletter or your own sales and marketing team's updated list, why not go for the opt-in before going in for the close? You need to date before you can marry, and email is one of the best relationship builders in the marketing universe.

Your collateral should support these parallel efforts and make it easy for prospects to learn more about your company through valuable emails. Caveat emptor: your email program can also reveal, for better or worse, company values, strengths, and weaknesses (meaning, do you bombard your subscribers with irrelevant offers or include typos, broken links, and other errors, or do you provide relevant and helpful information in a professional manner?).

5. Direct Mail

Sure, we all get bombarded by long letters with a special offer or coupons stuffed in a clear envelope, but this is missing the permission side of marketing, which also happens to be the secret sauce of any email marketing juggernaut. Direct mail can be a great bridge to costly, untargeted (and unsolicited) messages to a far more cost-effective, targeted, and permission-based marketing channel.

Remember, your names on direct mail are just lists; people that opt in to your company have given you permission. So, it's worth the extra step in transitioning potential and/or existing customers to that side of the marketing fence. Unfortunately, what may prevent this in many companies is politics, lack of cross-departmental synergies, or just not testing the concept out.

6. Mobile

What I call "Text2Grow" is beginning to take off, leveraging SMS' immediacy and efficiency for building an email program. While it may seem counterintuitive to use SMS to grow your email list, it's not. Most consumers are texting constantly with their friends these days, but not nearly as much with their favorite brands. Leveraging the ease of use of SMS for a quick consumer data entry point makes sense as an option. We had one client leverage this tactic on the jumbotron at a major sporting event where they were the primary sponsor. Talk about a high profile opportunity to convert a captive audience!

Don't forget your mobile site and/or smartphone applications when it comes to acquisition. I have downloaded hundreds of apps for my iPhone and iPad, yet have only seen a handful that provide me with the opportunity to engage with the brand on other platforms such as email, Facebook, and Twitter. This is a very real opportunity and one that's growing fast.

7. Experiential Marketing

So, what do you do after you walk away from a successful event-based marketing campaign (besides picking up trash and sleeping)? Whether it's a major sporting event or guerrilla-brand launch campaign in a pocket of a city, you're probably basking in the glow of a lift in brand awareness and product trials. That's all fine and good, but in the day and age of performance marketing, why not also offer your consumers who are being exposed to your product/service the ability to opt in to your email program and social presence? It can really blow away your ROI by stretching the exposure and relationship you began but never really built.

8. Catalogs and Packaging Promotional Inserts

We've all opened a package from a favorite catalog or website and found marketing fillers (Netflix trial anyone?), but rarely do companies cross-promote their own programs and additional services and offerings. Why not have a simple card or flyer placed in the package promoting ways to connect with the brand you just opened a package from? This could be your first opportunity to capture some people's attention if they are a gift recipient or a pure catalog shopper. Let them know they can "Like" you on Facebook or receive more special offers via email.

Most catalogs prefer that you receive their emails instead of the more expensive and often environmentally-frustrating, thick-as-a-brick paper option. Catalogs can merchandise their products and showcase items in different ways than email can, but ultimately, a customer who buys from emails is a far more profitable customer.

Pure e-commerce companies – don't forget your options. Even strictly e-commerce companies have relevant acquisition opportunities; some online and some offline. Call centers, transactional emails, customer service interactions, live chats, social networks, and any offline sponsorships and partnerships. The possibilities are significant as is the potential for monetizing these folks and generating an ROI to make any business swoon.

As is the case for capturing email subscribers on your website, your real estate placement, value proposition, and execution all matter greatly when it comes to propelling your email program's size on the back of your offline properties. Make it your summer plan to test out one area that's being underutilized. Your new subscribers will thank you.

Does $14 Million Worth of Email Addresses Cross the Privacy Line?

The disconnect between how executives and consumer privacy advocates view email marketing is a tenuous one. It was never more obvious than during the hijinks surrounding Barnes & Noble's acquisition of Borders' customer data, including email addresses in October 2011. It also makes for a great case study illustrating that the treatment of email addresses may seem trivial but it can be a huge liability as well as an opportunity to safe guard and showcase how you value your customer's data.

As part of the Borders bankruptcy proceedings, Barnes & Noble paid $13.9 million for Borders' intellectual property, including its 48-million customer database. That's a gold mine, or Pandora's box of data, depending on your view of this data. It appears Barnes & Noble got a dose of both.

Barnes & Noble execs who won the data in an auction (that phrase alone is enough to give many email industry folks cardiac arrest) thought the data was theirs to use as they saw fit, but were forced to reach a compromise, which meant Barnes & Noble was required to send a clear message to customers about its newly purchased data. Court-appointed consumer privacy ombudsman Michael St. Patrick Baxter became the watchdog in order to ensure the data transition to Barnes & Noble was handled appropriately.

The Wall Street Journal reported that Barnes & Noble fought the proactive opt-in approach every step of the way, including providing the proof of the proposed email to Baxter with a two-hour deadline to review (somewhere, email marketers are smiling at this familiar predicament). Barnes & Noble ignored Baxter's comments (again, hits home for email marketing pros), agreeing only to use his proposed subject line, "Important Information Regarding Your Borders Account," and one minor text edit.

Baxter, in court documents, said "It was clear to the ombudsman that a robust and meaningful opt-out was critical to reaching the negotiated privacy related terms of the sale. [Barnes & Noble's] failure to provide such relevant and material information in the opt-out notice may defeat the very purpose of the notice."

Borders' customers had until Oct. 29, 2011 to opt out of the Barnes & Noble email program. Otherwise, they would become part of the Barnes & Noble email program; and let me tell you, the frequency isn't light.

Some thoughts on how this situation played out and what could be learned:

- While other brand assets were included in the $13.9 million auction price, one could value the cost-per-email subscriber at $0.29. In a competitive industry like Barnes & Noble plays in, one could see why Barnes & Noble would be attracted to this deal. That is a very low acquisition price in any shape or form.

- Borders' subscribers and customers were likely book-buying consumers who may have found Barnes & Noble to be an attractive alternative and a relevant brand. But one would argue, if they had wanted emails from Barnes & Noble, they would have signed up for the Barnes and Noble email. Email permission is at its most powerful with an explicit opt-in and this was transferred permission that many viewed as unacceptable, unethical, or just creepy. Some viewed it as convenient. That is why...

- Execution matters. Clearly, Barnes & Noble executives wanted to preserve as much value in their $14 million investment and that would battle other concerns like privacy and opt-in versus opt-out. I am quite confident that the company probably didn't give too much thought on the actual message or execution of it. It was more likely, "It's our data and we are keeping it."

"I shall try to correct errors when shown to be errors, and I shall adopt new views so fast as they shall appear to be true views"
ABRAHAM LINCOLN

"Even if you are on the right track, you will get run over if you just stand there."
WILL ROGERS

CHAPTER 5
The New Best Practices

Oxford Dictionary defines best practice as "commercial or professional procedures that are accepted or prescribed as being correct or most effective." Herein lies part of the problem. Email as a new communication method and one that offers relatively easy ways to test and innovate should not be a medium that has "accepted procedures." This section is about pushing the envelope and exploring new ways to create value. Don't accept a best practice as what you should do – create your own best practices that are tried and true to *your* program and continue to expand them as your program evolves.

Do You Follow Email Marketing Profit Practices?

In a tweet[xxii], EmailMarketingReports.com's Mark Brownlow stated: "Wondering if 'best practices' were called 'profit practices' whether we'd be more likely to follow them. Words have power."

He's definitely on to something with that thought (among others). For many, the term "best practices" means old rules, processes, and methods to avoid if you are a risk taker or progressive thinker. Some let best practices restrict what they try to do (or even allow them to think further). And others use best practices to stay grounded and guide their email marketing decisions. I have heard some C-level execs say they aren't interested in best practices, and I don't blame them – at least some of the time.

So let's throw out the best practices rulebook and look at seven "profit practices" any email marketer should follow.

1. *Transactional emails rule.* Transactional emails are the easiest way to meet your customers where they are. If they've just abandoned their shopping cart or completed a form to receive additional information, they're obviously interested in what you have to offer – why not immediately send them an email to keep them coming back to your site or close the deal? According to "The Transactional Email Report" from Experian, transactional emails brought in revenues between about three and six times higher than bulk mailings from the same clients. If you're not sending transactional emails, you're missing a key profit opportunity.

2. *Respect permission based on lifetime value of subscribers.* If you know that an email subscriber on your list will spend hundreds or thousands of dollars with your company based on emails they receive, respect the permission they gave you. Do not bombard them with email or send irrelevant messages or you will risk losing that subscriber – and the revenue stream he or she brings

xxii http://twitter.com/MarkatEMR/status/25017565446

– forever. For the record, one study from Epsilon pegged the lifetime value of each email address at $23.

3. *Social media should be used as a profit generator, not a "gadget."* By using social media to obtain new email subscribers, you're giving your company a potential new revenue stream. According to ExactTarget's Facebook X-Factors study, 70 percent of customers who became a fan of a brand on Facebook did not feel they had given the company permission to market to them. Find a way to convert these fans to email subscribers who do give you permission – whether it's adding a link to your sign-up page on Facebook or building a new tab on your page. Don't just use social media to communicate to your fans; make it easy for them to become paying customers.

4. *For every email acquired properly, you have a better chance of making your numbers every quarter.* Email subscribers who have given you their explicit permission are worth more than those acquired through buying lists or other means – it's just that simple. Current and potential customers who want to hear from your company and are interested in your products are easier to convert than those who haven't given you permission. Why waste your time and resources sending emails to people who don't know about your company and don't care?

5. *The more you mail, the less your subscriber is worth.* Frequency is one of the biggest reasons that people don't read emails and unsubscribe from marketing lists. Hammering your list with offer after offer or deal after deal makes your recipients tired of hearing from you and less likely to purchase. Think of it like a teenager on a school break getting calls from his mom reminding him to get out of bed, clean his room, take out the trash, etc. – after the umpteenth call of the morning, he will stop answering the phone, go back to bed, and likely do nothing that Mom asked.

   In one period, Restoration Hardware sent me the exact same email seven (!) days in a row, slightly altering the subject line:

   1. A $100 Gift Certificate to Shop Our Fall Collection

2. Save $100 on Every $500 You Spend. Limited Time.

3. Save $100 on Every $500 You Spend. Includes Sale.

4. Save $100 on Every $500 You Spend. In Stores and Online.

5. Save $100 on Every $500 You Spend. Only 2 Days Left.

6. Last Day to Save $100 on Every $500 You Spend

7. Extended 1 Day Only: Save $100 on Every $500 You Spend

Guess what – I unsubscribed from the list due to the intense frequency . You had me at "Save $100" but lost me when you told me that seven times in a row.

6. *Invest in email marketing as a profit center, not marketing expense.* The ROI of email is widely touted as being the highest among any other marketing channel. Shouldn't you start treating your email program like the profit center that it is? Instead of thinking of email as a line item in the marketing budget, treat it as a revenue stream that should be carefully planned and nurtured. This means hiring the right team, investing in the right partner, and measuring beyond an expense line item.

7. *Content and value should guide your program, not marketing objectives.* This is the hardest of the seven to put into practice. Compelling content that provides value to your subscribers is the best way to ensure they stay engaged with your email program. It sounds simple enough, but finding a way to satisfy the marketing objectives while ensuring your emails provide value can be a huge challenge. Pay attention to and learn from your metrics – what are subscribers clicking on? What content prompts them to take action? Use that data to give subscribers what they want, and the ROI will undoubtedly follow.

Surefire Ways to Get Your Emails Read

While technology surrounding the typical email subscriber may be quickly evolving, the most important piece of the marketing equation will continue to be ensuring that your messages get read. If they don't, everything else is near worthless (outside of the branding implications that many hang their hat on). So how does a savvy digital marketer make sure his or her efforts aren't all for naught?

The From Line

The from line and subject line are the two most important influencers in getting a marketing email opened and read. Yet they are often given little thought, if any, during the campaign planning process. The from line is often the symbolic "who" for the other side of the subscriber permission agreement. As I have said before, if consumers opt in because you have a well-known brand, why in the world would anything be in the from line besides this recognizable and influential brand? It doesn't matter that your marketing manager set up the Email Service Provider (ESP) account or previously had their name in the from line. Unless you are in the B2B world and have proven that a personalized or well-known name in the from line works, stay away or at least test before removing your brand from this key space.

The Subject Line

Subject lines are a craft that often get little love in the email space. They often get the attention of your subscriber, or fail to do so, when they are unimaginative, tell too much, or don't stand out in a crowded inbox. My company tested using a hashtag (#FreeCookieDay) among the subject line and experienced higher open rates than normal due to a socially savvy subscriber audience and a unique element that separated us from the pack. Even if you are not a Twitter user, a hashtag (if it's clear or intriguing versus an obscure one) can be enough to stand out if your audience is a curious one.

Mobile email consumption has had a big impact on how consumers scan, read, delete and save emails. The subject line plays a big role in deciding your campaign's fate.

Personalization

Personalization of the subject line is one trick that even spammers employ but is worth considering in some situations. Don't forget any automated emails and their subject lines that may be triggered.

Length

Length is also important, and smartphones are changing the way and where people read emails, and many related strategies are increasingly worth reconsidering. We have clients who have short and to-the-point subject lines (think VIP invitations with clear and concise benefits/calls to action), and others who have lengthy subject lines that tease you into reading the email and acting. Bottom line: testing subject lines with specific regards to length can make a difference.

Why Your Brand Needs Email Smarts More Than Ever

Email marketing has been sneaky hot of late. In many ways it has experienced a healthy metamorphosing into part of a long-promised broader digital marketing arsenal focused on what matters most for the majority of businesses: making money. New and growing businesses are stealth in how they embrace email. You will hear marketing automation, cloud-based marketing platforms, digital messaging, CRM, and the like, but the companies valued highly and/or raising a ton of money or being acquired feature email front and center. I'm talking about ExactTarget, Pardot, Eloqua, HubSpot, Infusionsoft, and the list goes on.

These companies all showcase proven technology platforms designed to get your customers to buy more often and turn prospects into customers. So let's be clear. The technology side of the email business is deep and has a solid option for just about every type of business. What's missing still for the majority of brands is super smarts and know-how in terms of leveraging the technology or having a clear path to maximize ROI and (don't forget) give the best to subscribers that your brand can offer.

Anyone doubting the difference between a decent email program and a great email program can look to the Obama email marketing success during the 2012 campaign. This juggernaut blended email science and art to maximum success. No one can dispute the power and importance of online fundraising, and email was the bread and butter and just may have won Obama reelection. At the very least, it raised a boatload of money. Bloomberg Businessweek broke down its efforts and reported[xxiii], "Most of the $690 million Obama raised online came from fundraising emails."

This Obama email team had some serious smarts and results. Unlike the typical email program, they seemed to obsess over details, tested and measured everything, and then put these learnings to work. So they didn't do the typical "spend three weeks on an email and the last five minutes on the from and subject lines."

"The subject lines that worked best were things you might see in your in-box from other people," Toby Fallsgraff, the campaign's email director told Bloomberg. So forget long-winded themes or political messages – the best subject line was "Hey."

Email frequency was much discussed by campaigns as they blitzed their subscribers all election season. Their takeaway (and this may cause shudders to many email purists): no matter how many emails they sent, people wouldn't unsubscribe. Their metrics didn't show sending more emails resulted in any damage to the campaign. Certainly, this doesn't factor in emotionally unsubscribing, which is commonplace for email subscribers and their passive way of unsubscribing. But long-term customer value is relative when what you're selling has a limited shelf life.

It's unclear on whether sending more email had a direct correlation to raising more money, which seemed to be the ultimate goal of the

xxiii http://www.businessweek.com/articles/2012-11-29/the-science-behind-those-obama-campaign-e-mails email program (I would argue even more then getting President Obama reelected). I would bet more email resulted in more donations.

Other smart discoveries took a lot of people and time (Fallsgraff told Bloomberg that "we had 18 or 20 writers going at this stuff for as many hours a day as they could stay awake"). Things like "ugly" creative and unsophisticated tactics often won.

These findings, and more importantly successes, aren't the kind of thing that technology generates on its own. To commit to sweating the details, testing, and finding ways to make your emails more in line with what subscribers want and what your business needs is not an easy or simple fix. You must commit to your email program in an entirely different manner and be determined to build it out in a manner that accomplishes these ambitious but meaningful strategies and tactics. So as you seek a new email partner or build out a bigger and better team (remembering technology plus expertise is the first part of the equation), aim for bigger and better and don't fail to test or try something "just because."

B2B Email Marketing Tricks of the Trade

Business to business email marketing doesn't get nearly the attention as big consumer email programs but it is often where the most innovation is taking place. The following are eight ways to make a real impact on the bottom line. After all, that's what exceptional B2B email programs are all about.

1. *Email subscribers spend more than your other customers.* Well, they should. After you leverage email and marketing automation, you hopefully are touching your customers more frequently and in a more strategic fashion. We have helped most of our clients[xxiv] not only get incremental revenue from email subscribers but also the ability to measure it. Most CMOs and CFOs really like the measuring part.

xxiv http://brightwavemarketing.com/clients.php

2. *Ensure your creative doesn't suck – $1 million lift.* Most B2B email programs typically have a lot to be desired when it comes to creative. In fact, even smart B2B email campaigns I see fail on the creative. They are often long direct mail lookalikes that place the call to action at the email's bottom - the part that no one will read. B2B email creative isn't like B2C. In B2B, the goal isn't to sell the widget via email but to keep the relationship moving forward, stay top of mind, and differentiate your product or service.

   One of my agency's clients will see over one million in additional sales by optimizing templates that we made. These weren't just front-end tweaks to make it look cute. Changes were driven by better coding as well as data-driven decision making and testing.

   I like how Cisco WebEx uses its trial conversion series to highlight different features over the course of several emails rather than trying to cram all of them into one email. Their emails are clean yet provide a nice tip, some actionable items, as well as pricing and buying info. They included a casual note at the end ("BTW, have you downloaded the 'Getting Started Guide?'") which makes them seem less like an automated message and more like a personalized and (more importantly) helpful tip.

3. *Why the landing page is as important as the email itself (maybe more).* Ah, the landing page. Quite often it should be viewed as the no man's land of digital marketing. Many digital agencies let their rookie designer cut his teeth on their brand's dime. However, this approach can often make or break your email campaign and sometimes your B2B email program.

   Landing pages are wonderful vehicles when designed appropriately. They should be the continuation of the email's purpose and should offer value in exchange for a more specific relationship/sales opportunity. For example, tease the brand's new and exclusive white paper in the email and give it away on a well-designed and user-friendly landing page where more profile data might be acquired or, even better, a sales call might be arranged.

4. *Use segmentation and personalization.* One head of marketing recently told me that she considers her email program a precision marketing platform, and this is what B2B email marketing is when properly executed. Dynamic campaigns that customize the content based on a user's profile dynamically sent from a local sales rep (with the headshot varying based on each subscriber's profile) help bridge the gap of what I often hear many B2B campaigns complain of – being cold, too corporate, and bland.

5. *Data for remarketing.* One of the low-hanging fruits of the B2B email world is leveraging email metrics for follow-up or remarketing email campaigns. We have one large B2B client that saw a huge ROI due to applying the basic intelligence that email metrics offer up and using those to strategize a customized campaign effort that drove over $100,000 in incremental revenue.

   So in a nutshell, we took the response metrics and made a decision tree for the conversion goal and developed custom creative and messaging based on what the person did and what we wanted them to do. The efforts generated an ROI of almost 4,000 percent and won a plethora of awards from the digital and direct marketing communities.

6. *Use email to save money, not just make it.* We all know email is a revenue workhorse – in both the B2C and B2B world. Many B2B companies have realized that email can also reduce costs in other areas, making them even more effective. One client implemented an email program, not to sell, but to prevent expensive call center inquires. The email program saved the client over $1,000,000 in its first year. Not bad for a humble B2B email program.

   Too often, those potential savings are not communicated to other groups that could leverage them. Which leads me to...

7. *Don't silo your email program from the rest of the business.* In the B2B world, you must ensure you can fulfill each and every conversion. So if the email's job is to get the subscriber's attention and then move them through the sales funnel, the next step and certainly the conversion platform (which could be sales reps, call

centers, online, etc.) needs to be ready (not to mention aware of any campaign) and able to close.

We saw one B2B client increase its email-driven leads and revenue after adjusting its email campaign deployment times so call center representatives had adequate time to respond.

8. *The list matters.* This is often the tricky part. Too often B2B marketers fall into some traps. They buy bad data or spend their time hunting for a profile rather than permission-based leads. While I won't go into that contentious discussion right now, the quality of subscriber data is often the most important correlation to success. Meaning you knock my first seven tips out of the ballpark, but bad data can mean the first seven don't matter. So work on capturing relevant and accurate data versus fishing in a generic lead pile where the prospects don't know or care about you and your business.

   B2B email marketers are some of the best and brightest in the digital world – they just don't get the attention many other marketers do.

Email Reality Check

I speak at conferences often and run a lot of strategy sessions on email marketing with our clients as well as consult with financial companies investigating investments in the email space. These observations and takeaways are important for email marketers often living in their own bubble.

The Permission Division

While the power of permission is what makes email a true powerhouse in the digital world (without permission, email is a mere mortal form of direct marketing), the "what about this" scenarios that many ask show a big divide. Most marketers agree that "opting in" is the way to go but it is not often the reality in their businesses. The three-year-old sales lists, sister company email databases, and intriguing yet vague promotional names acquired are often too tempting not to add to their main lists or just use to "test the water." While many an email purist may not agree with this, it is the reality.

Relevancy versus Reality Question

Many of us preach the need to deliver relevant and compelling messages. Many email marketers in the trenches I talk to comment in some variation, "That's great but my customers and senior management don't care about relevancy – they want offers and something unique and valuable" (it's up for debate on who these email programs are being tailored toward). The interpretation of relevancy's true meaning may be in the eye of the inbox reader, but these email practitioners I often speak with think relevancy is, to a certain degree, a no man's land talked about by those not responsible for their own email program.

It's up to each brand to decide what is most relevant for their subscribers – remember, it's all about them not you!

Pencil These Questions In

It doesn't matter who, when, or where, I find these questions always get asked. No one can dismiss them since they continue to be at the forefront of marketers' minds. Some of them actually can define whether your emails get read as well.

- What is a good open rate?
- When should I send my emails?
- What makes a good subject line?
- What should be in my "from" line?
- How do I grow my list without spending any money?

I love talking to real-life email marketers who aspire for more in their programs and their subscribers' inbox. It's also equally interesting to talk with people on the outside looking in as their view of email is often very different and sometimes eye opening. Email marketers need to rethink the basics and the fundamental email issues to ensure we all improve – regardless of the latest and greatest threat/compliment to email marketing.

Email Tests to Try

Testing your email campaigns isn't easy. There's a lot that goes on: determining which pieces of the campaign to test, selecting message to test, finding time to execute the test, and analyzing results. In fact, according to Econsultancy's Email Marketing Census 2011, only one in three responding companies (32 percent) carry out regular testing for email marketing; one in four said they infrequently carry out testing, while 13 percent don't test at all.

If you test your campaigns, keep it up! If you're not testing, why not? Testing your campaigns can give you invaluable, actionable information to improve future campaigns immediately.

Starting today, pick one of these six components and test your emails to find ways you can improve for your clients and customers. It will likely pay big dividends in the future.

1. *From line.* The from line is the unsung superhero of every email. It's the first thing a recipient sees in his inbox, and it's the first impression the recipient has of your brand and message. Test your brand's name against a more personal connection (a sales rep's name, for example) to see which garners a higher open rate. But, be sure to avoid the random marketing person's name no one knows. One of my agency's clients tested personalization in their from line and achieved much more engagement than a generic from line – their sales reps' phones started ringing within minutes after the email was sent.

2. *Subject line.* This seems like a no-brainer, but it's often overlooked. Testing your subject line is easy to do and can provide interesting results to help guide your copy for future campaigns. Many in the industry think that having a short, specific subject line

with no mention of the brand and including only a message will provide the best results. While this may be the case, why not test it to be sure? Try testing long versus short, specific versus vague, brand versus no brand, etc. Use Google, Facebook, and LinkedIn ads to test the message that gets the most clicks.

3. *Mobile versions.* Experts from highly regarded organizations including Gartner[xxv], Nielsen[xxvi], and eMarketer[xxvii] all predict unprecedented growth in smartphone purchases and usage. You need to know what your audience sees on these devices when they read your emails. Develop a mobile version of your emails to accommodate your audience and their needs. For one of our clients, my agency developed both a desktop and a mobile version of the email that delivered basically the same message, but was easier to read for mobile users.

4. *Your data security.* With security breaches in the news across multiple industries during the past few years, it's more important than ever to ensure your email list is secure. Talk with your email partner to learn how they test their data security. Ensure their teams are doing everything they can to stay one step ahead of the spammers that would like nothing more than to have your customers' email address and other personal information.

5. *Opt-in form and process.* How easy is it for someone to sign up for your emails? Can they do it in one of your stores/restaurants/ locations? How soon after they sign up do they get an email from you? Ask your friends and family to go to your website and sign up for your emails, and then ask them about the experience: Was it easy for them to find the sign-up box or link on your home page? If it's not on your home page (it should be!), then was it easy for them to find? Or, if not online, was it easy for them to sign up in your locations? Is it in a logical place such as an "about us" or a news section? Did they know what they would receive even before they signed up? When they clicked "submit"

---

xxv http://www.gartner.com/newsroom/id/1550814

xxvi http://blog.nielsen.com/nielsenwire/online_mobile/android-soars-but-iphone-still-most-desired-as-smartphones-grab-25-of-u-s-mobile-market

xxvii http://www.emarketer.com/Article/Future-of-Smart-Mobile-Devices/1008228

or dropped their completed form in the box, how soon did they receive a welcome message? Getting answers to these and other questions about the email sign-up process can give you an objective peek into ways you can improve for your customers.

6. *SMS process.* When executed properly, SMS can be an easy, quick, and efficient way for your recipients to manage their email subscriptions or join your list. I tested the SMS unsubscribe process of 10 brands, and only one worked. While SMS isn't the be-all, end-all of email subscription management – it's clunky, includes costs for some recipients, etc. – it should be a more efficient way to get new subscribers or allow them to manage their subscriptions, including updating email addresses and unsubscribing.

When a Rose Isn't a Rose in Permission-Based Marketing

One afternoon, I was minding my own business when an email landed in my inbox and prompted me to do a double take. Not in a "Wow, that is something uniquely cool" kind of way. But in a "Legitimate brand just spammed me" kind of way.

Some may shrug their head and say "Duh." Still, I was surprised, frustrated, and wanted to try to make sense of this.

Here's what happened: I received two emails from two different brands within two minutes – RedEnvelope and ProFlowers. I sensed something was wrong.

I had previously opted in to receive email messages from RedEnvelope. While I rarely buy from RedEnvelope, its emails are generally well done. On the other hand, I haven't had any contact with or purchased anything from ProFlowers in years. I unsubscribed from email marketing lists for ProFlowers and other florists because they were sending out too many email messages.

As an email marketing purist and true believer in permission-based email marketing, I replied to RedEnvelope, asking why it transferred my email address to ProFlowers. I also replied to ProFlowers asking exactly how, when, and where I opted in to this list.

The good news? Each company quickly replied to my queries, which is often a surprise. The bad news? Representatives at both companies didn't fully understand the concept of permission marketing.

RedEnvelope responded:

> "I apologize for the confusion, however Proflowers is our sister company which is why you have received an email from them."

I hope you, dear reader, find this as unsettling as I did. After about 10 exchanges with customer service, one rep from RedEnvelope, Stephen S., provided the following explanation about how and why I was added to ProFlowers' database.

> "Basically Proflowers owns RedEnvelope. We are one in the same company operating in the same offices. No information was sold/given to any third party affiliate.
>
> ...RedEnvelope and Proflowers are the same company. We operate in the same database with the same customer base. Our customers know that RedEnvelope and Proflowers are the same company because of the logos that are displayed on each website. If you go to www.redenvelope.com, you can clearly see proflowers at the bottom where it reads, "Our Family of Brands." We even offer Proflowers and Shari's Berries items on the RedEnvelope site. Because we operate in the same building under the same management, no information has been given to any third party affiliate."

Indeed, San Diego-based Provide Commerce owns RedEnvelope and ProFlowers. However, none of that matters to me. Nor does it automatically transfer my permission from one company to the other. I didn't sign up with Provide Commerce (the parent) or ProFlowers (the sister), so the background family noise is irrelevant to me. I would assume 99 percent of their customers feel the same way.

The official response from Grace Lee in Provide Commerce's PR department was:

> "Through our emails we offer a wide range of gifting options, and great savings and deals that consumers might not otherwise receive. In our privacy policy we state that we may offer free electronic newsletters and promotional e-mails for products and/or services offered on any one or more of our sites, such as the email you received from a sister company of RedEnvelope. If anyone wishes to unsubscribe to our emails, there is an unsubscribe link at the top of the page, or they can call our Customer Service Department at anytime."

When I asked Lee if she thought this practice of non-permission was generally acceptable and in the customers' best interest, she replied:

> "Consumers give permission to receive emails from our family of brands through our privacy policy."

Most people don't consider the privacy policy to be part of opt-in; Provide Commerce is committing an egregious offense in my book. I wish I could say that this circumstance is a rare exception, but the sad reality is that many companies see this as fair game.

Customers typically don't know or care who the parent or sister company is. Unless I sign up with the corporate parent or a broad newsletter that outlines what I'll be receiving (and why), I better not get an email from another brand, company, or anything else affiliated with whom I provided my consent to for future mailings.

"These are customers! Why assume that they will want a sister brand?" said Stephanie Miller, VP, member relations at the Direct Marketing Association. "It's a great way to increase spam complaints and unsubscribe requests. It is OK to cross-promote. Why not send a great offer from RedEnvelope, featuring ProFlowers? That way, RedEnvelope serves its customers with cool, timely offers and adds value to the relationship."

A cross-promotional email would have been acceptable and probably relevant. But a full transfer of my email address from RedEnvelope to ProFlowers is unacceptable.

Did RedEnvelope violate CAN-SPAM? Nope. Did it cross the boundary of permission email marketing? Absolutely.

Bill McCloskey, founder of Email Data Source and Only Influencers, disagrees and offers an alternative view of this practice. (He analyzed ProFlowers' recent Valentine's Day campaigns and called it one of the more inventive and successful email marketers.)

"I think the definition of best practices is often misunderstood," McCloskey said. "Best practices should refer to the email practices that ultimately generate the most revenue, not some arbitrary code of conduct. If you spam people, your delivery will go down, you will be blacklisted, and you will make less revenue. In the case of cross promotion and sharing of sister company lists, the standard we need to look at is: does this practice increase or decrease revenue in the short run and in the long run."

It's not my intent to embarrass either RedEnvelope or ProFlowers. One of my goals in this book is to highlight a continued gap in email marketing – a battle between people who care about permission-based email marketing and those who "blast" emails to a "list." Humanize the experience and you will remember who is on the receiving end of these campaigns.

Advocates of permission-based email marketing must keep fighting the good fight. If we don't, we'll all just be sending emails instead of delivering relevant, valuable, and targeted messaging to people who asked for them.

The Best Time of Day to Send Your Emails Is...

What is the best time of day to send your emails? This question has been around since the dawn of commercial email. I thought about this some more, not to give an easy answer (10:37 a.m.), but because a lot of emails I was getting from brands I like seemed to come at less than optimal times, or at least for me and my busy inbox. Therein lies part of the challenge and dilemma of this topic. There is no magic bullet for timing, and if marketers think that their email will be lifted solely by the right time of day, that could create some disappointment. It deserves your attention though.

So let's look at some common mistakes on choosing the right time of day for sending your email campaigns:

- "Just getting it out the door"
- Pick the time that your boss wants
- Adapting to the "latest studies that say this is the best time"
- Disconnect between your subscribers, their email consumption, and your product/service
- Not looking at performance metrics
- Not testing

Below are thoughts on how to approach this in a different light:

- Put yourself in your subscribers' shoes. While this may seem like marketing common sense, I find this to be a gaping hole for email marketers. Internal pressures and thinking drive key marketing decisions. Choosing the deployment time often falls into this category.

  If you are sending an entertainment (e.g., dining, music, sports) related email, do you think your subscribers are thinking about where they are going to eat on Friday night on Tuesday mornings?

- Consistency and meeting expectations are key: a promotional campaign or B2B announcement featuring a new white paper are emails that can be approached in one way, as can a daily email or monthly newsletter. Of course, your email sign-up should spell out how often and (generally) when your subscribers should expect these emails. That's part of the permission agreement, right?

Gilt Groupe offers flash sales to its members, and its emails come within minutes of the sale opening. This makes sense for creating the sense of urgency in its offerings and is consistent with its business model. Inventory is limited and moves quickly, so that email is geared toward spurring quick action, not browsing and awareness.

- Asking your subscribers what time they want – I have heard for the past year or two that preference centers would become more robust and integrated with other marketing offers. Well, I am still waiting on that on many fronts and the amount of subscriber choices is still lacking. Why not ask your users when the best time to receive emails is for them?

- How is mobile impacting your subscribers' email habits? Almost half of every hour on the mobile Internet is spent[xxviii] on email. Chances are your email may be read on a mobile device. This has far-reaching implications, but at its essence it could mean your email will be deleted on a mobile device and not read in full until your subscriber is back at the house/office/preferred tablet reading spot.

- If you pass the mobile email triage stage (when a subscriber only reads/responds to necessary emails on their mobile device, deletes some, and keeps others to review on their preferred email consumption device), you are lucky. So the timing is really important in this regard. Is your business sending an announcement of a new software release at 8 p.m. PST when most of your subscribers are on the East Coast and probably read your emails during the work day? How about the daily deal that arrives early in the morning to a consumer audience that is

xxviii http://blog.nielsen.com/nielsenwire/wp-content/uploads/2010/08/us-mobile-time-spent-new.png probably focused on getting the kids to school and to the office? Neither sound like the ideal time. Moving a send time to a time that better synchs up with your product and users' potential behavior is worth exploring in more detail.

- What's the social angle? Social media has already received the obligatory best days and times of the week to post (Wednesday at 3 p.m. EST, says Vitrue). Of course, taking these with a grain of salt is required, but my company knows from our client work that on social networks there are spikes of activity that have regular patterns. It's different for each client, but keeping the social dialog activity top of mind should be a factor in sending your emails if part of your goal is to drive engagement on social networks. Either way, no one will share or discuss your email on social platforms if the content and value isn't there, so ensure that is the core of this social strategy.

While one of the golden rules of email marketing continues to be relevancy (and I believe a strong, value-filled email will be read and responded to no matter what day and time it is sent), digital marketers should seek to mitigate any risks and choose the best times of day for optimal response as well as peak user convenience.

Email Marketing Lessons From 2 Retailers

The holidays bring a flurry of email campaigns all fighting for your time, money and inbox attention. Frequency is the top issue for many, with most companies (especially retailers) pounding away at your inbox on a daily fashion (if not more than once a day); highlighting offers, free shipping, and whatever else you can say in 25 consecutive emails.The always innovative Zappos was one of the first retailers to let their subscribers guide the experience as opposed to being a part of the barrage whether you like it or not. Of course, unsubscribing is an option but many people want the email just not that many.

In an oldie but goodie, the shoe retailer took some bold steps in communicating expectations with its subscribers. Its email said holiday email frequency would increase, but only for five weeks. It more importantly allowed you to opt out temporarily or just receive one per month during this hectic time.

I really admire the transparency of telling subscribers how and when things will change. I don't have the inside scoop on how it reduced unsubscribes, raised sales, or achieved other goals, but kudos to Zappos for putting the subscriber first and being savvy enough to recognize the potential downside of a short-term email blitz.

On the other hand, as a customer and subscriber of Restoration Hardware, I felt as if I was about to be run over by its endless promises of unique deals and offers. In addition to its aggressive frequency, its emails were guilty of the dubious usage of "one giant image or bust." Generally, all of the email's value was in one giant image – disregarding the impact of image suppression (as most retailers unfortunately do). Well, that's the email geek in me, but the customer making some home renovations in me was more frustrated by failed promises and generally a lack of trust that came out of these emails.

It started in a promising fashion, with a subject line that grabbed my attention: "Reminder: Extended 1 Day Only. Save 20% During Friends & Family." I loved the feeling of exclusivity, as well as the seemingly no-strings-attached offer of 20 percent off everything. I didn't even notice the tiny asterisk that is omnipresent on most deals. Nevertheless, I forwarded it to my wife so she had the code and she happened to be in the mall that is home to our local Restoration Hardware. So, perfect viral case study in the making, right?

Well, apparently 20 percent off everything means only 20 percent off some things. So, a perfectly teed up email leads to a very unfavorable in-store user experience. Not the optimal outcome; and while email got the initial credit for driving forward a near in-store conversion, it also bore the brunt of our dissatisfaction, as I unsubscribed from future emails.

If your business depends on a seasonal home run, here are a few questions to ask (and not to ask) in your email brainstorming and planning sessions (you are having these, right?).

Do ask:

- How can we provide added value to our email subscribers?
- How will we ensure subscribers stay subscribers in the future?
- Why would someone sign up for our email program?
- What are our top three email marketing goals this year and how do they tie into our broader business goals?
- Do I have the right resources and partners to achieve my future goals?

Don't ask:

- What is a good open rate? A good open rate for me may be a poor open rate for you. Besides, the open rate is a loaded metric gun anyway (think preview panes, image suppression, and mobile rendering).
- What words should I avoid in my subject line? Instead of this question, focus on crafting an accurate and engaging subject line that provides a reason to open as well as a hint of what's inside.
- Should we do social media instead of email marketing? Instead, figure out how social media, or any other marketing channel, can complement and enhance your marketing programs, including email.

Spilled Ice Cream, Hurt Feelings, and Customer Choices

Let's just get the first part out of the way here. Ben & Jerry's never said it was eliminating[xxix] its email marketing program. Secondly, if it did, the world would go on and quite possibly the email marketing industry.

xxix http://www.bizreport.com/2010/07/ben-jerrys-drop-email-in-favor-of-social-media-marketing.html

Of course, this noise was enough for the email and social marketing space to furiously whip itself into a frenzied storm. We've heard and seen plenty of "Email is dead" headlines during the past decade, but this was a real live brand (and a beloved one at that) killing email, and it was almost entirely misconstrued.

A few years back, the U.K. operations of Ben & Jerry's announced that it would be eliminating its monthly email newsletter and moving its communications to Facebook and Twitter. Americans still hungry for flavored Ben & Jerry's email would not starve.

So, before we break down the moo that rocked a thousand inboxes that year, let's look at what Sean Greenwood of Ben & Jerry's told me when we traded emails about this:

"Email still plays a vital role in the Ben & Jerry's 'Marketing Mix' flavor! We have a great list of fans on our Chunkmail list that we send out notes to all the time with cool offers, promotions, new flavors & more. We're the luckiest company in the world because we have the best fans. After having actual fans suggest such powerhouse flavors like Chubby Hubby, Chunky Monkey and our #1 smash hit: Cherry Garcia, (all true!) who wouldn't want to be in touch MORE with our fans. Email correspondence allows us that opportunity. Plus everyone has their favorites: even Ben and Jerry! If email as a strategic tool helps us meet the need to provide our fans with choices, you won't see us cast it aside like an empty Boston Cream Pie pint container!"

The response to this news (accurate and not) was revealing. You had people questioning how and why you should shut down the pipeline that's going to deliver the highest ROI of any digital channel, some yelling who would buy ice cream because of an email or Facebook post, and some that drank the Kool-Aid without digesting the pint of facts that were being assembled in a quick-moving, free-for-all digital cage match.

So, rather than pontificating about which channel is the best and wondering if Ben & Jerry's U.K. team made a smart or short-sighted decision, I think it's best to look at this from a different point of view:

- Ask your customers what they want. Ben & Jerry's did what many marketers dare not do. It asked its customers directly what is best for them. In this case, what are their communication preferences?

- Give your customers what they want. Unless 100 percent of Ben & Jerry's audience said to replace email with Twitter/Facebook, then it is failing its customers. This rule is only for Ben & Jerry's U.K. team and customers (not to mention the social media prognosticators who failed to remove their blinders), but for anyone considering eliminating an opt-in marketing channel.

The Ben & Jerry's email/social hoopla continued to live on well after the fact – however, the company controlled the message, set the record straight, and did so via email to its subscribers. How deliciously smart and effective.

Is it wise to close any communication avenue where your customers and prospects have expressly told you they do want you to communicate with them through this channel?

The whole email versus social fight thankfully has been put to rest and the digitally savvy are focused on planning how to combine the two for the most powerful digital communication platter that a marketer could ask for. That's assuming they want to really combine performance, measurement, and engagement, and who wouldn't, right?

CMOs may have a one-track mind (and what's on that mind may be based on the day or the meeting), but every digital marketer needs to "get" that each digital communication channel is different and so are the reasons customers approach each one. According to research[xxx] from ExactTarget, people go to meet their favorite brands in the inbox for promotions and general updates, Facebook for entertainment, and Twitter to stay connected and "in the know."

---

xxx http://www.emarketer.com/Article/How-Email-Facebook-Twitter-Audiences-Different/1007829

"Consumers don't silo their engagement with brands to a single channel, instead they tend to 'layer' marketing channels on top of one another to meet their different objectives," according to the report.

With that in mind, should you kill a channel or bring a cut-and-paste messaging strategy to the table for your best advocates? Or do you listen, learn, and provide relevant and valuable messaging to each customer touch point?

Four Ways to Engage Email Non-Responders

We all know that many marketers view opens and clicks as the top two most important email metrics. I won't get into why this can be misleading or even irrelevant to your business goals, but we often consider unengaged subscribers as the ones that have not opened or clicked on a brand's emails for quite some time. This is an important and potentially confusing group of subscribers that have given you permission to email them but appear to be ignoring you and your message. Let's take a look at four ways to win them back and reengage them.

1. *Analyze the true data.* Taking an honest look at your list and diving into email domains (which can often provide insight into image suppression, which deflates opens rates), tenure as a subscriber, and buying history (or B2B actual lead status) can arm you with the right data, well beyond basic response metrics. After all, would you remove someone who buys regularly from you (or enough to be a profitable customer) based on the fact that they have not "opened an email" in six months? What if they have images turned off, only see your subject lines that trigger browsing through a different channel, or remember who you are and what you sell based on the emails they never read? These are the slippery slopes of making some decisions on who is truly unengaged and what determines this.

2. *Segment and customize.* After you have thoroughly determined who is inactive, there are several tactics that can potentially re-engage this audience. The first thing you must do is segment

these subscribers and suppress them from "normal emails" while they are in the inactive bucket.

- Subject lines. While you wouldn't scream at the person who ignores you at every Christmas party, you may change your approach. Subject lines are a great way to mix up your cadence, length, and approach.

- Creative. Maybe half of your list hasn't clicked on an email in the past six months because you send the same email with varying text or a boring template that never gets updated. Many of my clients have found a refresh of creative often generates new clicks from subscribers who otherwise have not given you a click. Of course, a mobilized version for many marketers is imperative these days and may be the reason a part of your database has not been active.

- Offer. Whether you are a B2C company hawking holiday widgets or a B2B company trying to grab budget before year end, mix up the call to action and value proposition. If you have a free research report, don't be shy about its placement in the email (too many B2B companies place it below the fold at the bottom of the email, old-school direct mail style).

- Frequency. While many would say to decrease the number of sends during your quest for reengagement, my agency has taken the other approach for some clients and have seen it work. Just like in person-to-person sales, sometimes staying in front of the prospect just works better.

3. *Automate a reengagement series.* Leveraging email automation based on business rules can be a very efficient way to have evergreen reengagement campaigns running once a subscriber is classified as "inactive." Don't forget to update these, as I have seen plenty of automated campaigns with a cross-promotion expiration date that is expired.

4. *Ask them.* I know it sounds shocking to ask these folks who have given you permission to email them, but seek further feedback

or even confirmation of whether they are interested in staying a subscriber, pausing their subscription, opting down (change frequency), or opting out. The last one is the one that gives heartburn to most VP/CMO-level execs since the size of the list is often one of the key barometers of the state of their email program.

What Can Email Marketers Learn From Lebron's Decision Fiasco & Other High Profile Backfiring PR Stunts?

Lebron James' prime time South Beach defection[xxxi] in the summer of 2010 was one of those moments that seemed to captivate America. I'm guessing Europeans mocked our culture and priorities, but it was a big deal stateside.

Well, Lebron's marketing and business team gets points for trying, although it looks like an air ball in hindsight. They crafted what seemed to be a unique strategic marketing plan that had win-win components and seemingly no downside. Delivering a huge audience for the top media dog in sports? Check. Raising awareness for Lebron's brand? Check. Including a children's charity? Check. All that was missing was an interview with Oprah but anyone that saw Lance Armstrong's post cheating admission, will know that doesn't always lead to a road of public forgiveness.

Imagine the feeling of sitting in a closed conference room plotting out such a good plan that no one has ever even dared to attempt it. Then you start to think of the rewards you will reap from this genius plan and how this will take your career to the next level. Or something like that before the door opens.

But then reality goes a different way and the reaction (externally and internally) isn't what you planned. Digital marketers can make this a teachable moment to prevent something similar from happening with their email program. Most of us can't withstand the negative impact

---

xxxi http://www.youtube.com/watch?v=RTeCc8jy7FI

that Lebron's image took, so what are the lessons learned here for mere mortals in the email marketing universe:

- *Know your customers.* Remember how and why the subscribers ended up on your list. If they signed up for a monthly missive from the CEO and your brilliant campaign involves a series of promotional offers, then you likely will and should fail. After all, your customers gave you permission for one thing and then you threw them a curve ball, which could alienate them much like most Clevelanders after Lebron left them high and dry for South Beach and the Miami Heat on decision night.

- *Have the business case to back up the risk.* Calculated risks are what often separate great business leaders from the average ones. What happens when you roll the dice on a progressive email campaign? If you are savvy, you have "sold" this internally to minimize the risk of exposure (yours and the company's). A home run makes you a star and a strikeout can permanently derail your ascension or be a minor hiccup.

  What makes the difference between the two? Usually a business case backed with metrics from a test (we all do this for each and every email campaign, right?) supporting demand, projecting results, and listing any variables that can impact success or cause a failure, as well as any contingency plans and secondary research that support this unique approach.

- *Control your message.* Any good public relations professional will tell you it's all about controlling the message rather than the message controlling you. If you go out on a limb with an email campaign, make sure you have the ability and thoughtfulness to envision how this may play out on the social networks and other channels that can help light a viral fire (for better or worse). Which means you must...

- *Have a plan B.* Any email marketer should plan for the worst just about every day. Most email marketers break into cold sweats thinking of the possibility of an errant email to a million subscribers, a suppression file going askew, or a subject line with an expletive. Whether by accident or a campaign that really missed the mark, a contingency plan is always needed. What do you do next? How do you say it or do you just hope for the best and stay silent?

Most digital disasters seem to happen when the silo is exposed and broken. Meaning, the liability with your email campaign may have nothing to do with your email program and team but on the tail of the click. If it involves a coupon, special giveaway, or something similar, ensure that you (and legal) pored over the terms and conditions, thought about the coupon blogs, photocopying, printing multiple, and so on and so forth. Basically all the stuff that could go wrong after the message is sent. If that isn't your group's responsibility, make sure you have assisted them with the due diligence, as it will be your problem should trouble surface.

- *Transition gracefully (and quietly).* Don't be mad at your subscribers for behaving in a way you didn't expect or in a fashion that disappoints you. All you have to do is go to any large brand's Facebook page to see even the most ardent fans act a bit unpredictable on occasion. So after you implement your contingency plan, adapt, monitor, learn, and move on.

Open Letter to Facebook (Per Your Email Program)

Always shocking to me (and maybe it shouldn't be) is when great brands that do amazing things on the digital and innovation front really stink at email. Those who phone it in despite having (seemingly) abundant resources and capabilities. Facebook is at the top of my list. This open letter is my plea to take email seriously and get with the times.

Dear Facebook,

We get that you may be worth close to $100 billion and that IPO and public company onboarding has been rocky but your future looks bright. Well, so does email marketing. I know you are busy, but I wanted to share a few metrics to convince you to properly invest in your email marketing efforts:

- Email is the preferred method of commercial communication by 74 percent of all online adults. Source: Merkle

- Facebook is the most commonly used social media site to be integrated into email campaigns, with 80 percent of North American online marketers having used it. Source: Lyris

- Nearly half of all daily deal subscribers were excited enough about them that they said they "can't wait" to see the latest deals in the messages. Sources: Yahoo Mail and Ipsos OTX MediaCT

- The vast majority of companies (72 percent) rate email as "excellent" or "good" for return on investment. Source: Econsultancy

- Email marketing generated an ROI of almost $40 Source: DMA

- 63 percent of mobile email users check their email account a minimum of once per day. Source: Merkle

- 93 percent of daily email users subscribed to marketing messages. Source: ExactTarget

I can let your COO's comment[xxxii] about email dying pass. I can also acknowledge that email is changing and that your Messages platform aimed to change it (although I think the jury is out and you failed). However, email is a huge part of what your one billion plus members do on your network. You can do better.

You slowly embraced email with a weekly Facebook page update and that was a decent start. Once you launched Deals and were diving into the daily deals space, you proved that you really got it that email will be the way for local advertisers to reach your members' inboxes. But guess what? To put it kindly, your email marketing stinks.

---

xxxii http://www.businessinsider.com/facebook-coo-email-is-probably-going-away-2010-6

In the interest of constructive criticism, I have offered up some advice (free – ends today!) below on how to get with the program and leverage this most powerful channel, not just use it as a necessary evil.

Creative needs some oomph. Your brand is a global and powerful one. Heck, you essentially had a blockbuster movie made about it. The emails, though, scream for more, and in my humble opinion look like they were developed by a couple of high school dudes in Word doc with an image (clip art of astroturf!) thrown in. Surely, you have the resources and assets to take it to the next level and beyond).

- *Calls to action.* OK, old-school email practices aren't your thing. But these emails cry for more calls to action, maybe another button, and something visually compelling that gets people to learn more and take advantage of these offers. Remember, links above the fold will be seen more.

- *Target.* This is necessary for better, more relevant offers. Wait a minute. Facebook, you know my birthday, where I work, and even what my kids look like. Surely you can deliver some more targeted and customized offers based on this treasure trove of data you keep on me and others, right? Let's use that for good, not just for conversations with Congress.

- *Give some love to the copy and messaging.* It also appears you are reusing the same pre-header for very different offers. "Buy a deal, invite your friends, and have new adventures" is pretty lame for a Keith Urban concert, Falcons game, and a local café offer tease. Yes, you really sent this email to me.

- *Add secondary and/or local content.* This could be the way to show that you have some skin in the game. This is a localized marketing effort and these emails (Atlanta deals, at least) sure scream that they were created on the left coast without boots (and care) on the ground in Georgia. Rival Scoutmob uses its quirky brand persona to demonstrate its local prowess while LivingSocial provides additional deals as opposed to just one in the event that it misses the mark.

Please know I will continue to read these – at least for another week or two as most of the offers prove to be off-target to me and I know you know a lot about me. You have a lot of competition and I want you to realize you have to be good at email marketing for this to work, regardless of your strengths in building community and offering cool functionality.

Yours truly,

Simms Jenkins

(former) Deals subscriber

In a Rut? Break Out Fast with These 15 Tips and Tricks

Email's neverending cycle brings fatigue. And frustration. Its lack of respect can compound these issues. So finding yourself (and your email program) in a rut can be common. The immediate and often efficient ways to improve your email program without committee meetings and department budget approvals is a good way to break out of your doldrums. Here are 15 ways to find some low hanging fruit and get reinspired.

Testing

1. Break out of the boring old look and feel. Knowing you get 10 seconds or less with your subscribers (if you're lucky), try to stand out with a horizontal email (see a gallery of some excellent side scrollers here[xxxiii]), test to see if that increases your response rate, and don't forget to nudge or cue your audience to follow the flow of the design.

2. Test some reactivation messaging for those pesky non-responders.

---

xxxiii http://stylecampaign.com/blog/2010/04/10-horizontal-scrolling-emails/

3. Recalibrate your email sign-up form and fight for some better placement to combat attrition.

Goals and Metric Challenges

4. Compare your year-to-date metrics to your original goals for the previous year.

5. Measure the value per subscriber and make sure your marketing leadership knows (because I bet they don't).

6. Ask your email partner what they think is working, what isn't, and what should be addressed to improve a few key metrics. If they're not telling you this anyway or, even worse, don't know, it's time to look for a new email partner.

Education

7. Pick 25 email leaders to follow on Twitter and LinkedIn (or their blogs, email newsletters, etc.).

8. Are you signed up for your competition's emails? How about your favorite brand's email promotions? Do you have a folder of your favorite emails or the elements you want to borrow from?

9. Have you joined the local interactive and marketing industry and networking groups or national ones like the Email Experience Council and Only Influencers? If so, make sure you're getting value from these and attending events. Why not even volunteer to speak or assist with an event on email?

Forward Thinking

10. Launch an email acquisition-focused effort leveraging your Facebook, Twitter, or LinkedIn presence.

11. Test video in a campaign.

12. Consider scrapping your email newsletter. No, not in the way the Ben & Jerry's U.K. team did, but evaluate the format that may be laborious to develop and painful for recipients to read. Try a new, simplified version that is built from the ground up with the subscriber in mind. That means, no one cares about the press release on the new Beijing office (unless your audience is China-targeted). Offer choices or deliver content based on click behavior and ensure it's not the same old content available on the home page of your site.

13. Integrate with social media. Your email message is not an island but rather a bridge which can lead to social media platforms. Don't be lazy and coast on a few simple buttons and calls to action. Truly try to use email as the vehicle for initiating a conversation that extends to a social media platform where your relationship is further strengthened. Social can thank email later.

14. Take a risk, any risk. Just so long...

    - As it is a calculated one
    - As it is one where you have a clear goal on what you're trying to achieve and how you will measure it
    - As it benefits the subscriber

15. Last but not least, thank and reward your subscribers for giving you permission to email them. You can never do this enough in my book (this one and the theoretical one). Remember, this isn't a media buy, so don't treat it that way. The best email programs are tailoring valuable information for this unique audience and treating subscribers like VIPs.

Pigskin and Email – A Winning Option

I live in Atlanta, GA. It doesn't matter what time of year it is but football is always king here. College and pro football are big (college undisputedly rules in the South)

The beginning of the season is filled with anticipation, hope, and lots of excitement (and, let's face it: beer). I have six rules that all football coaches know and email marketers can apply these lessons to their digital gridiron efforts (which often feels like an appropriate name for email marketing daily tasks). There are surprising parallels between football and email marketing. Use some of these to fire up your team and hit the ground running.

1. *Use the week ahead to prepare and test.* Anyone who has ever watched HBO's "Hard Knocks" or closely follows football, realizes coaching is hardly a cushy gig (despite the inflated salaries and requisite perks). In between navigating locker room feuds and team politics, late hours pored over game tape as well as preparing multiple scenarios demonstrate that success is often tied to preparation.

   Well, guess what – so is email. Your boss may think your email campaigns can be put together quickly, easily, and virtually for free (email is cheap and easy, right?), but the best campaigns have quietly been testing creative, messaging, subject lines, offers, and virtually every other detail that can impact success. Your emails may be sent every week, but just like any NFL or NCAA coach, merely showing up on game day without full preparation doesn't cut it.

2. *Defense wins championships.* Offense gets the attention and puts the points on the scoreboard. Promotional emails are the offensive juggernaut of any email program. Yep, they deliver revenue and what many of your customers want.

   However, just like a solid defense keeps your offense from having to do all the work and provides some much needed rest (ahem, frequency issues anyone?), newsletters and relationship marketing are the equivalent of a solid defense. Not as flashy as the offense/promotional emails, but they are what can build long-term success and be part of a combination that doesn't leave you a one-dimensional program. Don't forget, not all emails have to generate revenue to be successful. Emails that save money by providing a low-cost self-service option are great too.

3. *Block and tackle.* The hefty guys in the trenches clear holes for the offense and plug gaps on the defensive side of the house. Without them, you're not going to have a chance in succeeding.

   Design, messaging, and coding of your emails are the O and D line for your campaigns. Like offensive linemen, there are more rules in what you can't do than what you can do, and email designers have to abide by these protocols as well. Just because you are big doesn't mean you can block, and verbose copy certainly doesn't automatically translate to email success (in fact, it usually guarantees failure). Finally, coding an email can often bail you out of image suppression, mobile rendering, and more challenges. If you fail to address this area properly, your email will be sacked for a major loss, and that, my digital friends, can be painful.

4. *Go for the deep pass on occasion.* Email marketers are mostly a conservative bunch, but sometimes you have to take a risk on game day. While ensuring that each campaign advances in a strong but deliberate manner may be the goal of each campaign, every once in a while you should think about sending all of your receivers out for a deep pattern and let it fly. Take a risk every once in a while to keep your subscribers honest and see what happens.

5. *At halftime, don't be afraid to revamp the game plan.* Sometimes you may be on your way to a blowout victory – initial response rates through the roof and the champagne is on ice. Then, you start to coast and a quarter later you are feeling roughed up and the sideline anger is starting to brew. The boss calls an audible and you make some changes to that plan that seemed destined for greatness.

   Does it work? Does it matter? Every digital marketer needs more self examination and the ability to run the option and make that split decision and shift strategy on the fly. This versatility and nimbleness, after all, is a chief advantage of email versus its offline traditional brethren.

6. *Be prepared to be booed.* So after your campaign achieves record ROI and gets attention in the C-suite, will you get praise? Maybe,

but probably a brand manager or higher up will claim the victory and get the spoils. If an error occurs in the subject line or personalization goes askew, will you get called out in the huddle? You betcha. Be ready, and take the good with the bad.

---

32 Questions to Ask Your Email Partner

Whether you use an email-focused agency, an email service provider, or leverage in-house technology and services, how you use this resource is profoundly important in relation to your overall success. With big plans come tough questions, and very few successful email programs don't have at least one key partner. That partner may be a data specialist, offer technology solutions, design and code emails, or offer full-service email marketing.

Chances are, most are not delivering everything you need. Some may be mostly absent except during billing time and others may be your eyes and ears in the email minefield. However, they are working for you. So now is as good a time as any to probe deep and find out if this will be a lasting relationship.

Here are 32 questions to ask your email partner in order to maximize your relationship and ensure that they are the right partner or solution for your unique needs:

1. What do you do exactly for us each month and why shouldn't I try that in-house or look for another vendor?

2. What are the five best strategic improvements you suggested or assisted us with this year?

3. What are the five best tactical enhancements that you implemented for us this year?

4. How can our email program deliver more value to our customers?

5. How can our email program deliver more revenue to our company?

6. How will you help achieve both of these for us next year? Be specific, please.

7. What are you doing for me besides pressing the send button or providing a button for us to press?

8. Should I send more or less emails? Why?

9. What does your full service offering really look like and entail?

10. Am I paying more for technology or services? Should I do more or less of both?

11. Why haven't I heard from you that much, given all of the deliverability and inbox type changes (e.g., Gmail's Priority Inbox, etc.) in the email world?

12. Who should design my emails? How about coding? Should it be my ad agency, web shop, or you? Why?

13. What are you doing to differentiate our email program from competitors and make us stand out in a crowded inbox?

14. Who on my account team is proactively looking out for pitfalls and coming to the table with campaign concepts and strategic insight?

15. How well does my account team know my business and the email world?

16. If the CEO calls, how quickly can we get an important email out the door – from concept to deployment?

17. Tell me some success stories where you have done something innovative or special for a client?

18. We have other key channel partners including mobile and social – will you assist and play nice with them?

19. Social seems to be a big deal in the email world – how have you leveraged it for clients, not to mention us? What is the best social network for my brand to leverage?

20. Can you please explain your pricing and fees and help determine an ROI on this relationship?

21. What are your company's thought leadership initiatives, and do the people that work on my account participate?

22. I have to spend too much time making sense of the reporting – can you make it easier for me?

23. OK, I get it about our opens and clicks – but how has our email program contributed to our business goals and objectives? If it hasn't, why and what can we do to fix it?

24. How are you different from the other 37 email companies that I have heard from or about?

25. What makes your company special and a good fit for my needs?

26. Why do my emails that you "build" look the same as your other clients'?

27. Will you come to our office and give a C-suite state of the union on our email program?

28. Are you charging me for reading these questions and what other fees am I not aware of?

29. Why don't we talk or meet more often?

30. If you could fix three things about my email program, no questions asked, what would they be?

31. Will I get a better email partner if I switched to your competition? Why wouldn't I consider such a move?

32. What are you going to do to improve my email program next year?

Some of these may be shockingly basic, but many good email folks are lost searching for help. Use these for a straightforward request-for-proposal or just some talking points for your next partner meeting. Regardless, I have heard many people tell me they need more hands-on strategy and campaign management assistance but yet invest solely on the technology. That is akin to saying you are a bad driver but you solve it by buying a new sports car. Other times, an underachieving and incumbent email vendor remains as the partner because of inertia or a lack of incentive to find the best and, maybe more importantly, the right fit.

The buy-side of the email industry needs to dig their heels in and ask tough and potentially uncomfortable questions in order to get more value out of the email industry and ultimately help all of us move forward.

Why the Email Newsletter Deserves a Second Chance

Email's resurgence has been a powerful and yet subtle one. Many digital ninjas spout that email is dead and yet praise the business models of email-dependent companies, like the daily deal ventures (e.g., Groupon, Scoutmob, etc.)

With the focus once again back on the digital glue of online communications, email needs to continue to innovate. It can also return to basics for many marketers. One of those ways is the often shunned and regularly disdained email newsletter. I believe semantics have a lot to do with this prejudice.

The idea of newsletters conjures up mountains of text, hours of editing, and generally a painful experience for all involved with the creation.

Some of that may be true, but those are the elements of this communication piece that should be abandoned.

What I believe should get another look is an email product that doesn't even have to be called a newsletter. Let's call it a multi-dimensional email message. These are not promotional emails driving traffic to your site or store because of a sale on your widgets or a new white paper designed to build leads. These are emails that offer several unique messaging themes in short teases and then get the user on their way (often landing pages, websites, etc.).

Second That Emotion

A Nielsen Norman Group's research[xxxiv] on newsletter usability stated, "The most significant finding from our usability research on email newsletters is that users have emotional reactions to them. This is in strong contrast to research on website usability, where users are usually much more oriented toward functionality." This is huge and you should be sold on that nugget alone. It also sounds a lot like why social media works, right?

So if your brand wants to connect outside of a pure transactional and marketing context, then that insight makes it worth exploring; how you tap into that is the opportunity and challenge.

Multiple (Brief) Messages and Value Propositions

With email activity taking up almost 50 percent of every hour on smartphones, it's no surprise that the typical email newsletter consumption mode is skimming. So newsletters don't have to be the lengthy and boring version that many marketers continue to send (while often punishing themselves and their subscribers in the process) and users often ignore.

Make it brief and tease: Newsletter users spent an average of 51 seconds on each of the newsletters they read from their own inbox, according to Nielsen Norman's study. That's a considerably longer amount of time

---

xxxiv http://www.nngroup.com/reports/email-newsletter-design/

compared to a promotional email. However, that doesn't mean you need to hire a copywriter to write "War and Peace." Include short actionable items that appeal to multiple segments of your audience.

Provide varied messaging: A one-size-fits-all email doesn't work, especially in this format. So don't bet the farm that 100 percent of your audience is interested in one topic. This type of email allows you to provide multiple value propositions and types of content, which should mean low response rate, but a very diverse set of click-throughs that can provide valuable insight into what your subscribers find appealing.

Socialize the heck out of it: Content is often email's enemy, but guess what? It is social's best buddy, and email newsletters should be filled with solid content that should be created with social sharing in mind.

Develop some exclusive content: Yes, you will be repurposing content from other marketing initiatives, but include something that can only be found in your email newsletters (multi-dimensional email messages).

Personality Trumps Best Practices

Give it flavor and identity: Don't be boring with the newsletter. Give it a unique brand and personality, distinct from other email and marketing communications.

Personalize it: Ideally, subscribers can choose their type of content and frequency, but even without, a personalized email can go beyond Dear First Name. Customized content, either served dynamically or developed based on subscriber behavior, is guaranteed to be more relevant.

For instance, I receive a Dunder Mifflin email which offends me to the bone as an email marketer. It's copy-heavy, very few links are offered, and it looks like it was designed by my dad using Microsoft Word. However, I read it every time it comes. Not because of email bells and whistles, but because of my emotional connection to the funny folks from Scranton. Don't underestimate that potential.

When Automated Emails Attack

Demand generation has unleashed a whole new sector within the digital sales and marketing landscape. The industry is hot with lots of mergers and acquisition activity and in general interest and activity. Many of its principals love to take shots at email – despite the fact that email campaigns are the primary vehicle of marketing automation itself.

Demand generation is defined[xxxv] by Wikipedia as "the focus of targeted marketing programs to drive awareness and interest in a company's products and/or services. Commonly used in business to business, business to government, or longer sales cycle business to consumer sales cycles, demand generation involves multiple areas of marketing and is really the marriage of marketing programs coupled with a structured sales process."

To ensure a successful sales cycle, many marketers in this space have turned to email, and why not? Email helps to deliver the interactions the sales team needs and the information the recipients want. It serves as a key component of many demand generation/automated lead generation programs. And, it should be. However, it's not easy to develop a top-notch automated email program. Many of these are using the bullets in their gun the wrong way to their detriment. As a result, some demand generation marketers are accidentally shooting themselves in the foot. Here's a primer on what to avoid and how to optimize and better leverage the automated email piece to get more out of your demand generation programs.

Avoid the following:

Bad "from lines": There are different schools of thought, but I know I don't open (or notice) emails from people I don't know. Meaning an email from Richard Tyler doesn't get noticed by many busy B2B marketers. What if this poor, imaginary Richard guy is actually a smart and thoughtful sales rep from a brand you know and may want to do business with? Well, go with the brand in the from line without further

xxxv http://en.wikipedia.org/wiki/Demand_generation

consideration – it can go a long way toward getting Richard's foot in the door.

From lines are crucial in getting emails read and responded to, but I find from a survey of my own inbox that most B2B lead generation marketers use their name in the from line. Sorry, I don't know you and it may be spam rather than an opt-in and/or relevant email with some valuable content that could move me further along the sales cycle.

Harvesting and shady acquisition tactics: Let's face it – B2B marketers play by different rules than B2C. Many B2B marketers, including those in the demand generation/marketing automation space take some liberties in how they build their prospecting lists. In an ideal world (not to mention a CAN-SPAM-compliant one), email addresses would be all opt-in and no grey areas would exist related to permission. Some use social networks to grab contact info of contacts and prospects that have not opted in for their emails but may have a relationship with your company, while some go deeper and harvest names off websites and enter those in their funnel, which many would say is illegal and immoral.

Play it safe and smart: Use search, social, and your sales and marketing touchpoints to get permission – yes, it means creating some landing pages and being patient, but it is very well worth it in the end.

Misleading copy, long-form like letters, and shoddy creative: Somewhere, there must have been a misleading webinar. Many marketers are trying to fake you out or underestimate your intelligence with some emails (again, many may say simply spam) that either infer a relationship or conversation ("I am sorry we keep missing each other"). This is shady and can't possibly deliver meaningful results, regardless of what you are selling. Start off on the right foot and be transparent, no matter what your marketing message or channel may be.

At the same time, keep it brief. Marketing automation doesn't give you a free pass to overcommunicate just because it isn't a time-intensive "campaign." Too many automated emails suffer from brutally wordy copy that won't get read, and creative only someone in IT could love. Your automated emails need to have the same look, feel, and branding dynamics that your promotional email campaigns offer.

Lack of value and overly aggressive: Content is usually the star of these emails, much like the offer is in the B2C email world. If you have a great white paper, event, or something that will get past mobile triage, ensure it is clear above the fold, in the subject line, and in just a glance of the email. Remember, your audience isn't waiting for your emails – you are likely interrupting them so you need to offer a compelling reason for them to give you 10 seconds or so.

Don't go for the sale on the first (or second email) – this is a nurturing and relationship game here. Introduce yourself, why they should be interested, and what the value is for them with your company and these emails. Use the same rules as dating – start slow and build interest. Aggressiveness out of the gate often comes across as sleazy and will get you removed quickly, in both dating and in email.

Just because it is automated, sales-focused, or whatever other buzzwords that demand generation ninjas want to throw out, doesn't mean you should fail to bring your best email learnings and practices to the table.

"We are what we repeatedly do. Excellence, then, is not an act, but a habit."
ARISTOTLE

CHAPTER 6

Small & Medium Business Email Tips

Small and medium sized businesses and the entrepreneurs that run them are the heartbeat of our economy. Most businesses in these categories don't have an army of marketers nor the time to get educated on what makes email awesome or even some times what makes email not horrible. Most small businesses want cheap, efficient results when it comes to email. In this section, I will address some other key areas to consider when upgrading your email efforts.

The New Email Marketing Essentials for Small and Medium Business

I've had a lot of questions recently from small business owners and marketing directors from companies that are not enterprise-level brands. Many of them are related to the reality that email is changing and they don't want to be left behind. Or more importantly, their customers are changing in the way they interact with the business they buy from. They don't have the luxury of a big team or even smart, specialized agencies that live and breathe the nuances of email and digital marketing.

I have pulled together some of the most common and important questions that need to be addressed to ensure this vital part of our economy continues to use the right channels to deliver the right messages and keep their business profitable and thriving.

What is the real current state of email, especially when many peers and colleagues say they are moving to social media?

Email marketing and social media are both dynamic weapons in the digital marketing arsenal. Email and social media are both unique in their benefits and the tandem certainly share the power to engage customers and prospects for most businesses. Too much of the conversation these days is focused on which one is better and what is social media's impact on email marketing, but the reality of the situation suggests a different approach.

Email marketing's ROI in 2012 was around $40 (according to the Direct Marketing Association), which is far superior to that of most other marketing platforms. Social media, while generating much buzz and interest in the broader business community, falls short right now in terms of measurement, according to the big business executives. For example, General Motors pulled its advertising budget from Facebook, as it failed to see an impact on sales. Email's bread and butter is driving revenue in a targeted and measurable fashion.

How do I make them work together with limited resources?

The other part of the conversation is how complementary email marketing and social media are when planned and executed correctly. Email can often drive engagement on social media, and social media, while not done nearly enough, can help drive the more profitable email program as well. Chick-fil-A is a great example of a business that is integrating its email and social media efforts and letting its customers decide the way they want to interact with the brand while offering compelling opportunities to do so across multiple channels.

Too often companies operate all digital marketing efforts (not just email and social) in silos rather than focusing on how the customer may view and interact with the brand. Organizational silos can greatly impede your customer's experience rather than help it.

So the bottom line is that email and social media are two of the most dynamic and powerful ways to communicate with customers and prospects and can accomplish different things. I think if you ask any digital marketer the question, "You have one bullet left in the gun and need to drive revenue, what digital tool do you use?" you would find overwhelmingly the answer to be "email marketing." Of course, each company should be using multiple coordinated efforts to offer the best user experience with a specific business goal in mind.

Next, I will focus on some real methods that can help the small-to-medium-sized business to better leverage email when it comes to growing their database, how often they should send emails, and five essential tips that any sized business must be aware of when it comes to this powerful messaging channel.

Driving Your Small Business Forward With Email

Now let's focus on some specifics in other areas that any business owner needs to be aware of if they want to ensure email drives their business forward.

What are some tried-and-true methods that small-to-medium-sized businesses can use for their email marketing program and what are the big issues that we need to be aware of? Ones that don't require million-dollar budgets or a small army of experts?

List growth. The first and often most challenging thing small-to-medium-sized businesses need to do is develop and grow a true permission-based email subscriber list. There are no shortcuts and this requires patience. However, savvy businesses will capture email opt-ins at every customer touch point. Retail presences, call centers, their website, and social media platforms all represent perfect opportunities to capture email addresses in exchange for some kind of value. For some businesses it may be coupons and offers, while for a more B2B business it could be a white paper or valued content. Don't forget to send a welcome email and set expectations of what they should expect from your brand. Think about it the same way you would greet a possible customer walking through your store front or if they call your 1-800 number.

Frequency. How often you send emails is a crucial element to success and possibly, if abused, brand damage. A worst-case scenario is for you to gain trust and permission (and the potential for sales) via email sign-ups but due to excessive frequency, your recipients begin to have a negative impression of your brand.

Mobile. Smartphones and tablets are revolutionizing how people consume content and what, where, and why it matters. Nearly half of every hour on a smartphone is spent on email (Nielsen), so adapting your creative, coding, and strategy to this phenomenon is essential. Subject lines matter even more since boring ones or ones that don't entice may be deleted in line at the grocery store. For many brands, the goal is to not get deleted and to hopefully get read after the smartphone email triage happens.

Five other essential tips that businesses of all sizes need to remember for each and every campaign:

1. Customize and personalize your emails.

2. Use your metrics to guide future campaigns.

3. Be aware that email creative is very different than web creative.

4. If your brand is well known, use that as your from line – if a small business owner is well regarded, test using a personal name in the from line.

5. Make sure you know the rules of the CAN-SPAM Act – it's a federal law and has key considerations that many marketers and business owners are not aware of.

Sixty-eight percent of small businesses surveyed by Pitney Bowes listed email as their preferred marketing channel, so clearly email is the cornerstone of digital communication. A great, smart program moves the needle for businesses while providing value to its subscribers. The secret sauce can often be respecting the subscriber and sending relevant and valuable information, not just "blasting" emails when sales are slow. All of our inboxes are increasingly crowded, so to stand out and get read, marketers must apply common sense while leveraging the many possibilities of email marketing.

"Nothing great was ever achieved without enthusiasm."
RALPH WALDO EMERSON

CHAPTER 7

The People of Email Marketing

Anyone in the email marketing world or even those with just a little exposure to it, hopefully realize that the nicest and most genuine people work in our little part of the business world. While it is certainly a crowded and competitive industry, people are often there to help one another for the good of the industry. This is a field filled with hard working and humble people.

One of the best pieces of advice I can give is to reach out to some of these folks and ask for help and advice. It doesn't have to be a formal project or signed Statement of Work. Just stop and share what's working for you and how you got so many people to care about that great email you just sent. The email marketers of the world are focused on delivering strong results but also on extending a hand and their own knowledge base wherever possible.

This chapter is all about good people & good results. They do go hand in hand.

In Appreciation of Email Rock Stars

The day in the life of an email marketing manager can be a neverending series of challenges, requests, and navigating data and marketing mazes. All said and done, these marketing mavens and mavericks generally don't get a whole lot of appreciation or enough compensation[xxxvi] for the crucial tasks they perform and manage (25 percent of email marketing professionals said they make between $50,000 and $69,999 annually).

Let's look at a typical day for an email marketing program manager or digital director who handles the email channel, in order to gain better appreciation of the endless campaign cycle that is better known as the email marketing calendar. This is from the daily journal of Marty, the "interactive marketing guy who handles email" at a mid-sized brand.

8:04 a.m. – Check metrics from email campaign that deployed last evening

8:27 a.m. – Catch up on, what else: email

8:51 a.m. – Check tweets, industry articles, and peer discussion forums for the latest insights on best practices, case studies, and the like

9:00 a.m. – Meeting with IT to discuss automated emails that still contain random numbers in the "from line" and text that has not been updated for four years

9:45 a.m. – Coffee break

9:50 a.m. – Meeting with email marketing partner to discuss upcoming campaign creative, strategy, and execution

10:57 a.m. – Return voicemail from legal asking why we use logos and images in our emails

xxxvi http://www.emailstatcenter.com/Research/EmailStatCenter_CompensationAndResourcesStudy.pdf

11:11 a.m. – Work on campaign brief and strategy document for upcoming team meeting

11:57 a.m. – Lunch and learn digital meeting on social and mobile – no mention of how email will complement and integrate with these channels so need to set up meeting with these program managers

1:12 p.m. – Unexpected conversation with the CEO on the elevator – asked what I do again and told him my elevator pitch about our recent successes.

1:37 p.m. – Catch up on emails and competitive analysis

2:15 p.m. – Review reporting of email campaigns that deployed this morning

2:47 p.m. – Work with email partners on planning for monthly enhancements: reactivation campaign, new creative, mobile versions of email, list hygiene, testing subject lines, automated series focused on loyalty, Facebook acquisition efforts, new preference center, and about seven other things that will have to be addressed on the next call

4:14 p.m. – Production meeting to discuss proofs, edits, database segmenting, and related issues for series of seven campaigns that are being deployed tonight

5:04 p.m. – Meeting with boss about upcoming vacation (hoping hotel doesn't have Wi-Fi)

5:27 p.m. – Research and respond to email from department supervisor inquiring how much budget we can shave off email budget for the remainder of the year

5:39 p.m. – Research and respond to email from head of sales asking how much more revenue we can "get from more blasts"

5:50 p.m. – Coffee and break room talk with head of creative about "awesome online promotion launching this week." Was not aware of said promotion; plan on eating dinner at desk tonight

5:58 p.m. – Search daily deal emails for best (read: cheapest) delivery option since company won't approve meal expenses at work.

6:12 p.m. – Update email and digital calendar

6:27 p.m. – Catch up with deliverability guru on nasty issues that have surfaced with large ISP of late

6:45 p.m. – Send team weekly email brief letting them know of challenges with frequency and internal communication and need for more time and planning for promotional emails and the need for content for monthly newsletter that deploys in two days.

6:56 p.m. – Send note to web development team asking for better placement for email sign-up since it is not on home page and hard to find and new goal of 40 percent annual list growth will be difficult to achieve otherwise

7:26 p.m. – Review daily metrics, update production schedule, and can't help but notice job description that friend tweeted to me.

8:04 p.m. – Go home and shut off smartphone. At least until 2 a.m. when international email campaign is being sent

---

25 Things to Tell Your Friends and Family About Email Marketing

My three children are old enough to know daddy isn't a doctor, firefighter, or astronaut. So when they ask me to explain what I do for a living, I seem to always try a different approach. They know my company helps some of their favorite brands but they don't seem to understand how or why. When you think about it, we all need to take a step back and decide how do we answer those two questions: what do you do for a living and what does email marketing mean?

While many professions and industries have been around for decades, the email marketing world is new and it's important to define it. Not just to our families but to friends at cookouts, old classmates, and the

CEO with whom you occasionally share an elevator. So what does email marketing really accomplish? And what might be an appropriate answer for many of us, whether you are a VP for a large consumer company, in charge of lead generation for a B2B company, or an owner of a small business?

1. We help your favorite brands talk to their customers.
2. We can get urgent bulletins and news to interested parties in a flash.
3. We send special messages wherever you are – work, home and on the go.
4. We provide relevant information where people spend most of their time – when on their computers, checking email.
5. An email from a favorite brand can make you smile, save some big bucks, or teach you something you didn't know before it landed in your inbox.
6. We can make the public keep talking about your Super Bowl ad six weeks after the big game.
7. When an email goes out, it can make your sales teams' phones ring.
8. Our emails provide benefits that other marketing channels can't match.
9. Email marketing empowers my best customers and provides my company with rich insight into what customers are interested in and what they are not interested in.
10. When an email promotion goes out, people start spending.
11. We let our customers and prospects know when we have something really cool to share with them before the general public.

12. We provide a crucial connection between our customers and our brand.

13. Our email newsletter lets interested parties know what is new with our company.

14. We save our company money by communicating more efficiently.

15. We provide real benefits to our best customers that they choose to receive.

16. We drive interest and traffic to our sites and stores.

17. Our emails help our employees across the world know the latest and greatest news.

18. We deliver breaking news to where you want it.

19. When an email offer goes out, thousands of people feel special.

20. We keep our customers loyal and coming back.

21. When we have a new product or service, we use email to tell our audience about it.

22. Email marketing is our most valuable, targeted, and measurable form of communication.

23. We deliver what people ask for.

24. Our customers say they love our emails and the beneficial information and offers they get each month.

25. We provide the strongest return on investment within our department.

Some of these can hopefully assist you in articulating what email does best, whoever the audience is. Maybe it can provide the outside world clarity on what email does and help open a few eyes within your company.

Who knows, maybe one of these will end up in an industry campaign like "Got Milk?"

---

7 Habits of Highly Effective Email Marketing Leaders

Has your email program made serious progress this year? How do real email marketing leaders avoid complacency? What makes up a great email marketing maestro? Let's evaluate:

1. *Constantly update your executive team.* I don't mean sending 24 slide reviews on each campaign or a spreadsheet with nine tabs. The best email leaders can demonstrate their effectiveness (and challenges) in a few bullets or in their 30-second update during a meeting. Think big picture impact, and as I have said many times, your CMO doesn't care about opens – tell them about conversations, conversions, and revenue.

2. *Reinforce goals.* I have walked into many meetings with clients, prospects, and email managers and the first question I often ask is "What is your email program trying to accomplish?" If there is a pause or not a clear consensus, then the team is not clear and this is a problem. Too often, the person responsible for the day-to-day has a goal of getting campaign X out the door rather than a broader goal.

3. *Provide clarity on subscriber value.* We all know what email programs we love and ignore. The best ones, we already have a connection to the brand, so a big chunk of the hard work is already done. Now, you have to deliver consistent and clear value to the subscribers. An email leader with a big picture view can accurately sum up why someone should sign up for their emails and what they will get.

4. *Hire and monitor the right team.* This could be the most crucial item, and means internal and external hires. Should you really

hand the keys to an inexperienced internal person? Do you hire a generalist marketing agency (digital or traditional) that treats email like an afterthought and has the track record to match? Do you solely rely on a technology platform that doesn't provide the ROI you need without the right services and hands-on expertise? You must be able to answer these questions or success will likely be unattainable. Don't forget, it's your job to challenge them and don't let the program run on autopilot even when assembling the right team.

5. *Challenge conventional thinking and best practices.* An email marketing pro with a bright future is going to push the envelope, test what works (and what doesn't), and not accept the answer "It is a best practice so that is why we are doing it."

6. *Coordinate and integrate.* The right email marketing leaders really are digital experts and they just happen to be a subject matter expert. Anyone who knows email and only email has a one-way ticket to eternal middle management. The true email visionaries are plotting their moves to enhance social, mobile, offline, and any other channel that touches your customer. Don't get caught in the email off-ramp without having the digital road map.

7. *Be passionate and articulate.* I have seen plenty of talented digital practitioners get left behind because of being swallowed by organizational inertia, not fighting the "Email is cheap and easy" syndrome, or lacking the ability to defend why email is a channel that commands the appropriate resources and attention. Stand up, state your case, and watch your career soar alongside your email ROI.

---

25 Things Email Marketers Must Avoid

After the first few weeks of the year, do you fall back into bad habits or the email hamster routine that allows for little else other than tactical management of the next email campaign (and there is always the next email campaign)? Or do you create big strategic initiatives and a laundry list of plans to evaluate for the coming year? Getting them done is hard. Ignoring them is easy.

My recommendation for each new year: have a list of big picture initiatives and blocking-and-tackling measures. Then figure out the best way and resources to accomplish both of them and establish a reasonable timeline. Make sure you avoid some traps that strangle progress and end any chance for positive momentum. Avoid the following pitfalls and continue to navigate the path to success. Overlook them and it could be an uphill battle.

1. Little to no exposure to the C-suite.

2. Talking about your program exclusively in opens and clicks.

3. Working with a partner or platform that has not elevated your email program at all.

4. A welcome message or series that was created by IT six years ago and hasn't been updated since.

5. No new acquisition tactics to grow your list and minimize subscriber churn.

6. Lack of focus or coordinated efforts on how the social program can help email, not just the other way around.

7. Not dealing with deliverability and list hygiene issues, assuming that problem will just work itself out.

8. Knowing that your typical email campaign looks subpar on a smartphone and tablet but not understanding why or doing anything to address this.

9. Refusing to revisit campaign processes and the parts of the campaign development that take away valuable time and resources.

10. Improving your subscriber sign-up experience at all touch points (website, retail, mobile, social, landing page, etc.).

11. Assuming frequency has no correlation to the success and monetization of your email program.

12. Telling your subscribers way too much, all in the body of an email, as opposed to a brief and compelling email that moves them through to the next point.

13. Subject lines that suck.

14. Not auditing your CAN-SPAM compliance and unsubscribe process beyond the link at the bottom of your email.

15. Accepting inferior creative designed for non-email usage.

16. Treating all of your subscribers equally.

17. Not having exploratory conversations with some of the cutting-edge vendors that touch email and have recently emerged.

18. Sending the exact same message more than once.

19. Not doing any A/B testing.

20. Assuming best practices are the status quo and can't be tested and challenged.

21. Not asking more questions about why an email campaign is being sent and clarifying the business goals and rationales of each campaign.

22. Not finding time to network, talk shop, and learn from peers in the trenches.

23. Ignoring what your competitors are doing via the email channel.

24. Keeping email in a silo and not bringing other internal groups to planning and operational meetings.

25. Not tracking ROI or some kind of "killer metric" that will demonstrate the effectiveness of your email program as well as the job you are doing in driving these important efforts forward.

"Good business leaders create a vision, articulate the vision, passionately own the vision, and relentlessly drive it to completion."
JACK WELCH

CHAPTER 8

The Business of Email Marketing

I have touted for years the crucial success factors of having executive buy-in and being able to articulate the business purpose of the email program. This is a chicken-and-egg scenario, but get 10 email marketers in a room and they can talk about their open and click targets and list size goals. Ask them what that means for their business and you may be in trouble.

As digital marketing evolves from niche channels with small budgets and low expectations to a fundamental part of most businesses, the ability to tie your email program to your overall business goals is of paramount importance. I tell my clients and internal team to just imagine being in the elevator with the brand CEO and when she asks why are we doing email and what is it doing for our business, you need to have the answer – and a good one. Remember, CEOs don't care about opens and clicks.

Email marketing is often a tough sell to non-believers. Whether Wall Street or a high school buddy, email marketers often get dismissed as spammers or old fashioned irrelevant internet marketers. Of course, if you didn't already know, then hopefully this book has further convinced you that email is one of the biggest engines of the digital economy. We often have a hard time articulating this to our senior management, clients or naysayers and this hurts our overall reputation and long term progress. This chapter puts a spotlight on how to measure and present our strengths and some big picture email success stories.

How Much Are You and Your Email Program Worth?

One of the most persistent conundrums of email marketing is its incredibly high ROI[xxxvii], yet it is often underfunded and understaffed. What's more, those managing email marketing programs are undercompensated. EmailStatCenter.com, an email metrics portal my company founded with the Email Experience Council, set out to take a deeper dive into how email programs really looked under the hood in 2010. What we found is concerning, though not totally surprising.

In the Compensation & Resources Report[xxxviii], we surveyed over 200 email marketing professionals on the client and services (agency, ESP, consultants) side.

The Client-Side Challenge: Resources and Budget

Over 40 percent of client respondents stated they had $100,000 or less of their annual budget dedicated to email marketing

- Fifteen percent had $100,001 to $249,999
- Fourteen percent had a budget of over $1,000,000
- Additionally, fourteen percent did not know their budget
- Thirty-seven percent of client-side respondents said they have only one to two people within their organization who are directly working on email marketing

---

xxxvii http://www.emailstatcenter.com/ROI.html

xxxviii http://www.emailstatcenter.com/Research/EmailStatCenter_CompensationAndResourcesStudy.pdf

- Thirty-four percent said they had three to five on their team

While budget is always a clash in the email world, larger companies (22 percent surveyed are companies of 1,000 or more employees) will have an uphill battle on maximizing their email programs on a budget south of $100,000 and with only a few people on the team. A major cause for concern: 14 percent did not know their budgets and almost 40 percent have one to two people working on these efforts. While this may speak to the often inexperienced teams managing email programs, it also highlights the need for managers involved in this essential channel to better understand the broader business goals and restrictions they may be facing.

Service-Side Dilemmas: Spread Thin in Offerings and Accounts

Despite email being a highly specialized area of digital marketing, it seems those working in the service-side of email may be spreading themselves thin.

Other lines of business that email service firms offer:

- Strategy and consulting (66 percent)
- General interactive marketing (53 percent)
- Social media (43 percent)
- Web design (42 percent)
- Search (34 percent)
- Mobile (30 percent)

Traditional advertising firms appear to be flat-footed in offering email, despite its near universal adoption as a marketing channel. Only 36 percent say they offer email marketing. It is also worth noting that 37 percent of survey respondents on the service side said they work on 11 or more accounts, further supporting that service-side email firms need to increase staffing.

Compensation in the Email Marketing Industry

Here you may find the ammunition to get a raise or conversely, you may find that you want to keep this data to yourself:

- Twenty-five percent surveyed said they make between $50,000 to $69,999 annually
- The next highest bucket was $35,001 to $49,999, closely followed by $70,000 to $84,999
- Less than 5 percent of participants make $200,000 or greater per year

Interestingly, marketers on the agency side earn higher salaries than their client-side counterparts. However, those at the director level and above tend to earn higher salaries working on the client side.

Our findings showed that the size of the dedicated team correlates with the overall email marketing budget. Teams that run the majority of their email marketing efforts in-house must staff their teams with more experienced, thus higher salaried, employees. The median income for employees managing programs with only one or two dedicated email marketing resources ranges from $50,000 to $69,999 for companies with an email marketing budget of less than $100,000. For companies with an email marketing budget of more than $100,000, the median salary ranges from $70,000 to $84,999.

You can take these findings with a grain of salt or you can use them to help fund and staff your email program to get the most from this highly targeted and measurable channel. The choice is yours.

Email's Place in the Emerging Entrepreneurial Economy

Email has certainly cemented its place as the digital marketing backbone. Sometimes many professionals living the email marketing "dream" aren't aware of the role their sector is playing in the larger economy. Too many email marketers get defensive in the "we are not spammers" mindset and forget to wake up and smell the ROI. The same goes for other marketers who largely ignore or take for granted email's role in acquiring, monetizing, and retaining customers.

With that in mind, I wanted to take a look at the companies recently recognized for their high growth and job creation, as cited by the prestigious Inc. 500|5000 list[xxxix].

The 5,000 companies on the 2012 list reported that they created over 400,000 jobs in the past three years, and aggregate revenue among the honorees reached $299 billion. So I took a look at the email-related companies on the list, at least determined by me as their primary focus of business, which is very subjective and may not be 100 percent accurate.

There are certainly other recognizable technology/marketing hybrids like Pardot (no. 172) and HubSpot (no. 314) that have email as an offering but again, I went with email as their primary line of business. I found six email-centric firms in the mix (including my own company, BrightWave Marketing).

The companies and their Inc. 500|5000 listing:

Bronto – No. 1272

BrightWave Marketing – No. 1335

Emailvision – No. 1397

ExactTarget – No. 1576

ClickMail Marketing – No. 1992

EmailDirect – No. 2370

VerticalResponse – No. 2802

---

xxxix http://www.inc.com/inc5000/list/2012

These six companies had approximately $332,100,000 in combined 2011 sales and created almost 1,300 jobs during the previous three years (according to Inc. statistics).

There are also several email-focused public companies that deserve significant recognition. Many may paint themselves as multi-channel and cloud computing firms but the ones I listed are most known for their email platforms. ExactTarget, like Facebook, finds itself on the Inc. list and a newly minted publicly traded company.

ExactTarget: Employs over 1,100 and is worth $1.61 billion.

> Responsys: With almost 800 employees, the California company has a market cap of approximately $333 Million.

Constant Contact: Has over 900 employees and a market cap of just under $480 million.

Lyris: Listed on the over-the-counter market (also known as "pink sheets"), the company has a market cap under $20 million, which surely is much less substantial than many of its privately held peers in the email marketing world.

(Note: All market caps listed as of February 15, 2013)

So what is the takeaway here? Well, for any of the doubters still questioning email's longevity and greater influence in the economy, I hope this list provides real numbers and a real place among the digital elite for its job and wealth creation, not to mention larger impact on the communities and businesses it serves.

---

8 Things I Wish Everyone Knew About Email Marketing

I was reading marketing pioneer Seth Godin's blog entry[xl] called "8 things I wish everyone knew about email," and as Mr. Godin often does,

xl http://sethgodin.typepad.com/seths_blog/2010/04/8-things-i-wish-everyone-knew-about-email.html it got me to dig a little deeper. What did I wish everyone knew about email marketing? Remember, there is a huge difference between email and email marketing. The result, for better or worse, is this list.

1. Email marketing has a huge ROI[xli] but it isn't automatic. In fact, it takes a lot of hard work, smarts, and creativity to achieve this kind of return on investment. Oh, you have to work hard to track and measure it as well. If you do have a return on investment that would make your CFO happy, be sure to let the CFO know this important fact. Too often email programs (the great and poor ones) stay off the radar.

2. Email creative is very different in almost every element than any other marketing creative, including other digital mediums. Businesses often treat email as an afterthought, using it to drive traffic to their spiffy microsites. Guess what, it shows.

   I asked a former colleague with a creative bend to summarize this situation and he told me, "Email doesn't support a lot of the 'bells and whistles' you'd find in more traditional digital media (advanced interactivity, animation, sophisticated/consistent rendering), therefore you have to be even more creative to still produce a dynamic and engaging design." Which leads us to the next thing I wish everyone knew about.

3. Your emails don't show up looking the exact same way for all of your users. Images don't magically show up for everyone either. In fact, Marketing Sherpa research says[xlii] that only 33 percent have images turned on by default.

4. CAN-SPAM (and other international spam laws) aside, email marketing occurs when permission is granted. Spam is when there is no opt-in. I don't think permission differs from B2C to B2B and I believe email and business rules don't change either based on who you are sending to.

---

xli http://www.emailstatcenter.com/ROI.html

xlii http://www.clickz.com/clickz/column/1716214/disabled-images-email-disable-interest

5. Email marketing managers really want and need management's approval (just like everyone else in the business world who feels occasionally marginalized or taken for granted). The chief marketing officers of the world should embrace email beyond its quick and easy revenue stream mentality and see it (and its practitioners) as a long-term digital bridge to your best customers.

6. Even your best customers don't read really, really long emails. Length does matter. The job of most emails is to get your attention and solidify that interest and relationship and drive you somewhere else, whether that is a landing page, retail store, or white paper library.

7. Email marketing and marketers want to be your friend. Email marketing should be the go-to complementary tool for whatever sales and marketing effort you have in the pipeline. Need traffic to your new website? Email can assist. Can't get above a few Twitter followers on that corporate account? Enlist email to demonstrate why someone should follow you. Email is the digital glue for any company, so ensure that you consider it well before any campaign launch.

8. To grow your list, you have to make it easily accessible at any customer touch point, make the process seamless, and offer a compelling value proposition. I haven't seen many companies grow their email database in a significant fashion through burying their forms, failing to provide a clear idea of what a user would get should they sign up, or generally any reason to opt in.

---

5 Ways to Sell Your Email Program to the C-Suite

I'm having a lot of interesting conversations with clients and prospective ones excited about being on the verge of something big. It might be a major investment in the program, a new partner, or internal recognition after years of hard work. I think most in the email space have that feeling too – we are onto something bigger than ever before and the timing is right to seize this opportunity.

However, digital marketers often feel they won't be able to make "the leap" – and not because of execution, customer adoption, or anything related to their core brand and its strategic benefits. What lurks on the other side is more of an internal problem tied to a general fear that someone will not understand the full scope of why and what's so important. After all, most email programs work pretty well even when poorly planned and executed. So these smart and savvy marketers I have been chatting with want to ensure they pitch their program properly to have it "blessed" by senior management and usually more importantly, not squashed by these same people.

Here are some practical ways to get buy-in from the C-suite:

1. *Find the right key performance indicator.* Sometimes any business-related goal tied to your email program can be enough, but you might as well go for the right one, not just any metric. Connecting your email program's impact to a key performance indicator (KPI) like revenue-per-subscriber or sales-per-campaign will all of a sudden make your email program stand up strong next to its less measurable digital cousins like social.

2. *Have a state of the union meeting.* Email often doesn't get much of a spotlight because it is hard to shine when its leaders are hiding under a rock. Often it is because of the Sisyphean tasks, but other times it is because email managers make false assumptions of what internal teams know about the program or don't give themselves enough credit in what they have accomplished.

   Change this with inviting all of marketing (or in a smaller company, the entire team) to listen to what the email program has achieved and where it is going. You might be shocked at the response you get. Even if you can't get on people's schedules, always have this deck ready and be sure to update it monthly.

3. *Start communicating differently.* Stop forwarding emails with long and hard-to-decipher analysis and spreadsheets of campaign performance. Do you think the CMO reads that? We arm many of our clients with a high-level scorecard that connects the email program to the rest of the business. This is what you

report if stuck in the elevator with the CEO and he asks how your program is doing.

4. *Creative sells.* While creative is one of the many weapons in an email arsenal (sometimes self-destructing in the wrong hands), let's face it: people like to see and touch things and even more so if it is a pretty picture. Email momentum has slowed many a time when the business rationale failed to have a tangible example of how it was manifested to the customer. Show killer creative and you'll help your case in a meaningful way. Throw in a mobile version of some campaigns and you may be perceived as cutting edge.

5. *Business cases, not theories, get investment.* Ultimately, you want to sell your program not just for a pat on the back or a raise, but to grow the program, try new things, and drive the business forward in a stronger fashion. So besides summarizing what your program does and why subscribers have embraced it, you need to be able to succinctly articulate (think one slide, not 20) what your program is capable of doing with additional support, resources, and/or investment. Meaning don't go down the path of how a mobile preference center will increase your subscription base by 10 percent due to increased smartphone adoption by a large segment of your best customers. You lost your CFO early in that statement.

Project that "our email program will contribute $2 million more in revenue (or whatever business metric you can estimate) by leveraging new tactics that correlate to changing consumer habits." A recent study showed 89.6 million Americans used their mobile phone to access email during a three-month period (comScore) and one way we can monetize this is to create ways to interact with these customers where they are spending more time."

Remember, you often can't improve on many fronts if your program doesn't get broader visibility. This is one of the biggest challenges email marketers will face internally and it must be addressed if you have the ambitions to take your program to the next level.

Connecting in an Expanded Messaging Marketplace

I consider ExactTarget's wildly popular conference Connections to be a great barometer of not just where email and digital messaging are in the industry but where they are going and want to be. I know I'm not alone as many brand-side marketers attend despite not having any relationship with the publicly traded email and messaging platform provider.

At one recent conference, I expected to see some big themes that reemphasized the continual need (for revenue, public relations, and strategic needs) for companies like ExactTarget to move beyond the inbox. Here's what I believe to be the key agenda items for similar conferences over the next few years.

Bridging the cross-channel gap. I believe this to be a bigger issue and not just one CMOs say they want and need in forward-thinking meetings. Email programs need to be better connected to other channels and vice versa. Too often strategies are not in the same ballpark for complementary channels (think social and email), yet they work best when synchronized.

To me there has never been a better time for customer and subscriber growth opportunities – you just need to figure out what to leverage, how, and why.

Making sense of mobile. Yes, the past three to four years have seemingly been the "year of mobile," but it's time for digital marketers to actually figure out what mobile means for them. For some it's apps and mobile ads, for others it's SMS and emails on smartphones. Well, it's all of the above in my book, but I know marketers are seeking to find out how mobile fits into their digital messaging efforts.

For many, I hope this is the push they need to build and deliver emails ready for tablets and smartphones and offer consumers choices on how and what type of messages they receive. Regardless, mobile will be on everyone's mind as consumers expect a superior mobilized experience, and many marketers are stuck in 2005 in developing strategies, campaigns, and experiences for their customers and prospects.

The future of email is headed where? This orange-flavored event is often one of the most influential to a considerable percentage of attendees and many want to come away with a vision of where the industry is headed as well as their careers.

Well, ExactTarget does a good job of guiding many minds. In fact, it did so in the fall of 2012 with the announcement of two acquisitions[xliii]. Integrating two different types of businesses into an expanding suite of digital tools is what many vendors seem to think marketers want. ExactTarget splashed into this space years ago with the purchase of CoTweet, an important strategic move in telling the world (especially Wall Street), "We don't just do email." So getting ahead of the curve can be as important to being there when many potential trends become reality, and ExactTarget has made some big bets on this.

What Needs to Change in Email Marketing

The first month of 2010 provided more email marketing industry merger and acquisition action than in the past two years (based on my unscientific tracking). In Q1, we saw boatloads of new studies demonstrating email as being the primary driver (and beneficiary) of social media.

Email is finally getting the praise, attention, and money it deserves, right? The big breakthrough we've all been waiting for has occurred – or has it?

Research from the Society of Digital Agencies (SoDA) found[xliv] that 34 percent of senior marketers regarded email marketing as a low priority. Only "games" was ranked lower among the nine digital categories.

---

xliii http://phx.corporate-ir.net/phoenix.zhtml?c=218491&p=irol-newsArticle&ID=1744377&highlight=

xliv http://www.emarketer.com/Article/Engagement-on-Social-Networks-Top-Priority-Marketers/1007479

Clearly, email has made a big leap during the recession and this new research counters other findings that show email as a top choice of all marketing channels. However, it demonstrates that email often isn't on the forefront of the digital radar for CMOs as much as we may believe. It can and should be though.

For that reality to take shape, several big and small things need to fall into place. Among them:

Leadership

The technology side of email marketing (think ESPs) has rightly led email on many fronts. However, in the long term that can't be the only way. Too many email programs remain stagnant and woefully under-perform. This is partially based on a typical email scenario. Company A has a few interactive generalists, one manages the email program, and their only partner is an ESP that is really selling a technology platform.

The ESPs get paid based on volume, not success. The technology and fees related are tied to a platform that sends messages. Obviously, this is a crucial element to any email program, but one that will not generate success by itself. Therefore, more companies and leaders need to emerge on the other side of the fence.

Historically, most of the thought-leadership has come from the technology side, and that has been and will continue to be invaluable. However, the client side and the email-focused agencies and specialized firms that one could assume have more influence over an email program's success need to step up. Certainly, there are a few top-tier email agencies and some vocal client-side advocates of the channel who have been doing this for years, but a shift should change in order to get the focus back on how to optimize your email efforts, not just technology bells and whistles that don't get used, or the spray-and-pray strategy of sending more emails.

Hopefully the email community goes for bigger wins in this decade and doesn't focus as much on the insider pet projects (spelling, nomenclature, and other things that fuel email's Napoleon complex) that the outside digital community doesn't care about. Let's demonstrate the bigger digital picture and our essential role in it.

Change the Cheap and Easy Notion

The highly fragmented nature of email marketing has created a confusing maze of companies offering seemingly similar products, services, and golden tickets to high ROI. With that have come viable low-cost, self-serve tools and cheap monthly pricing. A boon to small businesses, yet noise in the back of the typical CFO's head when he sees a line item more than four figures for email marketing.

"I thought email was cheap?" goes the typical refrain from the unfamiliar executive. With that, we must continue to explain why it's worth investing strategically in email marketing and building the business case and models that demonstrate the significant impact email can make on the bottom line. The old notion of, "To make money, you need to spend money" rings true for most in the email world.

The "easy" side of the email coin has a lot of the same fundamentals. To the untrained and unfamiliar eye, email marketing success looks teasingly simple. Take some images from offline advertising, plug it into a tool, and fire away to a list accumulated over the years.

Quickly, once someone digs deeper, a world that includes changing best practices, CAN-SPAM and related legislation, testing, list hygiene and proper acquisition tactics, analytics, segmentation, image suppression, rendering, and deliverability issues lurks below the surface. Until the C-suite understands that email deserves a proper team, tools, and partners, it will be hard to take the next big leap.

The Bandwagon Has Room

You don't hear much from the people who said that email was dead, but you sure hear from the folks jumping on the email bandwagon these days. Maybe it's due to the CFO-friendly ROI of the channel, but email has dropped its old-school shackles and become the belle of the ball again.

Yep, email is not just hot in a marketing kind of way, but in a central foundation business model kind of way. DailyCandy blazed trails, Groupon set them on fire, and now even the humble email newsletter is the platform of choice[xlv] for hot-to-trot entrepreneurs looking to distance themselves from the crowded social/web 3.0 catfight.

As tech entrepreneur Jason Calacanis tweeted: "Everyone is rocking email now! :)"

Think about it – if you have a brand, product, and/or service, something (content, offers, a new product) an audience might like to hear, and the wherewithal to see it through, you too can launch an email marketing-fueled company. Let's look at some things that prove that email is the digital channel to beat:

- The investment/financial community has noticed. This goes well beyond Groupon's halo and the other companies using email as their "product." The companies that provide the services and technology to these firms, not to mention the brands that have long relied on email to communicate to their customers and prospects for years, are experiencing major growth.

  The M&A space is hot in this sector and among my email marketing brethren, one has even just filed for an IPO[xlvi]. Expect a lot of transactions in the email service and technology space to continue throughout this decade, which will keep email on the front page of the business section.

- Much like search in the early 2000s, every hot digital channel needs a wingman. Search needed email to form a bridge to talk to these newly acquired customers and leads. Social needs email to drive the conversations and traffic to these new engaging destinations and then monetize them.

xlv http://techcrunch.com/2011/01/03/digg-founder-kevin-rose-launches-private-newsletter-called-foundation/

xlvi http://www.bizjournals.com/sanfrancisco/news/2010/12/23/responsys-files-for-60-million-ipo.html

- Mobile has become huge in many ways (the iPhone is single-handedly creating new business models thanks to the success of its apps) and not as much as expected in others (SMS opt-ins to brands have not grown in the way most predicted), but it doesn't matter. Mobile won't even describe one of the ways people use the Internet in a few years, as it will be a de facto way for exploring anything and everything online. In fact, mobile traffic grew 10 times since the year end 2009 and mobile advertising has grown at a compound annual rate of 153%, according to venture capitalist Mary Meeker.

  We all know people are using their mobile device for a plethora of tasks, and in multiple locations (meetings, grocery stores, dinner tables, etc). But what are they doing? Almost half of every hour on the mobile Internet is spent[xlvii] on email, by far more than any other task (yes, that means you social networking). So get religious with mobile email and capitalize on location-based email consumption, as most of your subscribers are reading your emails on the go – or it may be ugly.

- Email automation can kill your dinner or yourself. Email automation and lead cycle programs are hot, and rightly so. A well-planned-and-executed automated email campaign that provides a nurturing and helpful path for your prospects can be incredibly rewarding on many fronts.

  The reality is that most have great intentions but suffer from horrible creative and content (five direct mail type letters won't get read, I promise!). Additionally, they are often plagued with high frequency and an internal marketing-driven path rather than demonstrating clear value for your subscriber. Do these right or they will backfire.

- Email creative is sexy! Responsive design, mobilized email, HTML 5 video for email, subject lines with cool images in them

---

xlvii http://blog.nielsen.com/nielsenwire/wp-content/uploads/2010/08/us-mobile-time-spent-new.png (think hearts and airplanes), cinemagraphs, animated gifs and more bring email to new heights to get your attention.

Bottom line, email will remain the digital hub in the near future as it has withstood several potential knockout punches in the last 10 years. It is never too late to embrace its power, reap its rewards and lead the charge for more relevance, greater innovation and deeper relationships. You and your subscribers will thank you later.

Conclusion

The new inbox may be literally crowded, filled with new icons and types of messages and to some a hated ritual of cleansing one's communications. To others, it is a place filled with anticipated offers, heartfelt missives and the first place to check as one wakes up each morning. Metaphorically, it is a special and sacred place for those that are invited in, but often viewed as the cause of a love-hate relationship. Those that invade its space are treated like the scourge of the planet. Those that deliver the goods have their stock go up in the minds of the consumer (sometimes even literally).

This is why marketers must treat their email and messaging efforts with more care, respect and trepidation. The subscriber invites the brand to communicate with them and that is unique, special and valuable. It's different than blanketing a city with billboards and radio ads to hopefully reach a tiny part of a possibly interested marketplace. When you grant permission to a brand, you are expressing some kind of interest. There is an unspoken contract here. That's the power of the new inbox.

There is of course huge opportunity that comes along with this for both sides. Finding the common win is what makes the new inbox so elusive and inviting. Good luck in getting there.

SPECIAL BONUS

15 Email Marketing Statistics to Impress Your Boss

1. Both marketers (76%) and consumers (69%) favor email as their first online "check" of the day. - ExactTarget (2013) "Marketers from Mars"

2. "Mobile purchasing decisions are most influenced by Emails from companies (71%) only surpassed by the influence of Friends (87%)." Adobe

3. 75% reported they would resent a brand after being bombarded by emails. - Emailvision (2013), "Survey Reveals Bombarding Consumers with Marketing Results in Brand Resentment"

4. Marketers who take advantage of automation—which includes everything from cart abandonment programs to birthday emails—have seen conversion rates as high as 50%. - eMarketer (2013), "Email Marketing Benchmarks: Key Data, Trends and Metrics"

5. eMarketer estimates, the mobile internet audience will reach 115.8 million by year's end, comprising 38.5% of the population. By 2016, that number will increase to 60.5%. - eMarketer (2013), "Email Marketing Benchmarks: Key Data, Trends and Metrics"

6. 77% of those surveyed reported that email was the preferred channel of communications for promotional offers. - ExactTarget (2013), "2012 Channel Preference Survey - Report"

7. 50% felt getting their name wrong was a reason to think less of the brand. - Emailvision (2013), "Survey Reveals Bombarding Consumers with Marketing Results in Brand Resentment"

8. Top ten leaderboard of the most popular webmail, desktop, and mobile email clients: #1 Apple iPhone 23% (down -0.41), #2 Outlook 16% (down -0.43), #3 Apple iPad 12% (up +0.17), #4 Apple Mail 9% (down -0.2), #5 Google Android 8% (up +1.15), #6 Live Hotmail 7% (up -0.72), #7 Yahoo! Mail 7% (up +0.02), #8 Gmail 5% (up +0.42), #9 Windows Live Mail 3% (down -0.03), #10 Yahoo! Mail Classic 1% (up +0.14). - Email Client Market Share (2013), "Email Client Market Share Leaderboard"

9. 33% of consumers believe companies should invest more of their marketing time and resources to improve customer loyalty. - ExactTarget (2013), "Marketers from Mars"

10. 63% of consumers prefer email as a channel to share content with friends and family. - ExactTarget (2013), "2012 Channel Preference Survey - Report"

11. Email marketing Return on Investment (ROI) for 2012 is predicted to be $39.40 for every dollar spent, which will account for $67.8 billion in sales - Direct Marketing Association (2011), "Power of Direct"

12. Nearly a third of survey respondents (32%) said they have stopped doing business with at least one company altogether as a result of its poor email practices. – Merkle (2009) "View from the Inbox"

13. By 2016 $1.5 billion of the projected $2.5 billion total spend on email marketing will go to agency focused disciplines like creative, integration and analytics. – Forrester (2012)

14. 93% of US online consumers are email subscribers, receiving at least one permission-based email per day. – ExactTarget (2011) "The Social Breakup"

15. Email marketing remains critical to business, with 89% of respondents declaring email to be "important" or "very important" to their organization. - DMA (2013) "National Client Email Report"

For more email marketing metrics, please visit EmailStatCenter.com

About the Author

Simms Jenkins is CEO of BrightWave Marketing, North America's leading email marketing focused digital agency. The award-winning firm specializes in elevating email marketing and digital messaging programs that drive revenue, cut costs and build relationships. Jenkins has led BrightWave Marketing in establishing a world-class client list including Affiliated Computer Service (A Xerox Company), Chick-fil-A, Cox Business, Phillips66 and Southern Company. The agency was recently ranked among the fastest growing private companies by Inc. Magazine.

Jenkins was awarded the prestigious AMY 2010 Marketer of the Year from the American Marketing Association for being the top agency marketer and the Email Marketer of the Year at the Tech Marketing Awards held by the Technology Association of Georgia. Jenkins is regarded as one of the leading experts in the email marketing industry and is regularly cited by the media as such and called upon by the financial community to provide market insight and consulting.

Jenkins is the author of *The Truth About Email Marketing*, which was published by Pearson's Financial Times Press and is currently the Email Marketing Best Practices Columnist for ClickZ, the largest resource of interactive marketing news and commentary in the world, online or off. His industry articles have been called one of the top 21 information sources for email marketers. He has been featured in Fortune Magazine, The Wall Street Journal, Adweek, Bloomberg TV, Wired Magazine and scores of other leading publications and media outlets.

Additionally, Jenkins is the creator of EmailStatCenter.com and SocialStatCenter.com, the leading authorities on email and social media metrics. Prior to founding BrightWave Marketing, Jenkins headed the CRM group at Cox Interactive Media.

Jenkins serves on the eMarketing Association's Board of Advisors among other civic and professional boards. He is also a mentor at Flashpoint, a Georgia Tech based startup accelerator program. Jenkins is a graduate of

Denison University in Granville, Ohio and resides in Atlanta's Buckhead neighborhood with his wife and three children.

Follow and connect with Simms on Twitter (@SimmsJenkins), LinkedIn (http://www.linkedin.com/in/simmsjenkins), and his book websites at NewInboxBook.com and SimmsJenkins.com.

39412616R00095

Made in the USA
San Bernardino, CA
25 September 2016